*Practical Statistics
for Engineers and Scientists*

PRACTICAL STATISTICS FOR ENGINEERS AND SCIENTISTS

NICHOLAS P. CHEREMISINOFF

TECHNOMIC
PUBLISHING CO., INC.

LANCASTER · BASEL

Published in the Western Hemisphere by
Technomic Publishing Company, Inc.
851 New Holland Avenue
Box 3535
Lancaster, Pennsylvania 17604 U.S.A.

Distributed in the Rest of the World by
Technomic Publishing AG

Printed in the United States of America
10 9 8 7 6 5 4 3 2

Main entry under title:
   Practical Statistics for Engineers and Scientists

A Technomic Publishing Company book
Bibliography: p. 201
Includes index p. 205

Library of Congress Card No. 86-72352
ISBN No. 87762-505-0

TABLE OF CONTENTS

PREFACE

This monograph is intended as a handy reference of standard statistical analyses and data regression techniques. It is written in a short format to enable the user to obtain a working knowledge of important statistical tests. Each method is fully illustrated with practical problems drawn from engineering and science-oriented applications. The volume does not provide rigorous treatment of statistical theorems and theory, but lists useful references for those readers who desire more in-depth understanding of the mathematical bases. The book also provides direction in constructing regression routines that can be used with worksheet software on personal computers. A software program called DATA-FIT developed for simple linear and non-linear regressions and capable of handling up to three variables is described in this volume. The program is designed to run with LOTUS 1-2-3 spreadsheet on an IBM compatible personal computer. Appendix A provides a full description of the program and can serve as the user's guide. A floppy disc of this user friendly program can be obtained by using the convenient tear-out order form in the back of this volume. In addition, the volume provides a listing of various commercial software programs and the statistical analyses capabilities associated with each.

Science and engineering students should find this to be a good supplemental reference in course work and in graduate studies. Process engineers, chemists and researchers involved in experimental programs and pilot plant operations may find this to be a handy guide in designing experiments and analyzing data.

Special thanks are extended to the staff of Technomic Publishing Co. for the production of this volume, and their continued interest and attention.

NICHOLAS P. CHEREMISINOFF

ABOUT THE AUTHOR

Nicholas P. Cheremisinoff heads the Product Development Group of the Elastomers Technology Division of Exxon Chemical Co. in Linden, New Jersey. Among his areas of interest are multiphase flows and rheological and processing behavior of polymeric materials. He is the author and co-author of many engineering textbooks, and is a member of the AIChE, Tau Beta Pi and Sigma Xi. Dr. Cheremisinoff received his B.S., M.S. and Ph.D. degrees in chemical engineering from Clarkson College of Technology.

CHAPTER 1

Notations and Definitions

USE OF SUBSCRIPTS

Subscripts are used to denote the number of an observation in a set. For Example: x_i is the ith observation in a set. $x_{(i)}$ is the ith ordered observation in a set. Let us review Illustration 1.

Illustration 1

Consider the following data set $\{5, 10, 6, 2\}$. The observations are:

$$x_1 = 5, \quad x_2 = 10, \quad x_3 = 6, \ x_4 = 2$$

$$x_{(1)} = 2, \quad x_{(2)} = 5, \quad x_{(3)} = 6, \quad x_{(4)} = 10$$

Summations are denoted by the symbol of capital Sigma (Σ) and the operation is performed over the limits of the observation.

Illustration 2

$$\sum_{i=1}^{4} x_i = x_1 + x_2 + x_3 + x_4$$

$$\sum_{i=1}^{n} x_1 = x_1 + x_2 + x_3 + \ldots + x_n$$

$$\sum_{i=1}^{n} x_1^2 = x_1^2 + x_2^2 + x_3^2 + \ldots + x_n^2$$

$$\sum_{i=1}^{n} cx_i = c \sum_{i=1}^{n} x_1 = c(x_1 + x_2 + \ldots + x_n), \text{ where } c = \text{constant}$$

$$\sum_{i=1}^{n} c = nc$$

1

$$\sum_{i=1}^{n} (x_i \pm y_i) = \sum_{i=1}^{n} x_i \pm \sum_{i=1}^{n} y_i$$

$$\sum_{i=1}^{3} \sum_{j=1}^{2} x_{ij} = \sum_{i=1}^{3} (x_{i1} + x_{i2}) = x_{11} + x_{21} + x_{31} + x_{12} + x_{22} + x_{32}$$

DEFINITIONS OF DISTRIBUTIONS

An important step in analyzing the statistical significance of data and in regression is the layout and tabulation of data values. *Frequency distributions* are a useful format for scanning and reviewing the important features of large bodies of data. In preparing frequency distributions one must select data groups. Important rules-of-thumb to apply when choosing data groups are:

- Find the minimum and maximum values in the set.
- Do not use fewer than 6 or more than 15 groups.
- Select unambiguous grouping criteria.
- Group intervals should be of equal length.
- Select a convenient group midpoint.

An example of preparing frequency distributions using the above rules is shown below.

Illustration 3

Bales of rubber are checked in a warehouse for brown spots. These brown spots are considered as contamination and can be the basis for rejection by customers. The number of brown spots per bale is tabulated in Table 1-1 for 100 data sets.

The minimum and maximum values in the set are identified in Table 1-1. We

Table 1-1. Data Tabulation of Brown Spots.

6	20	6	20	26	15	12	19	20	30
3	23	3	(33)	15	0	23	7	19	8
10	19	7	14	20	18	24	8	22	13
20	17	7	15	18	17	25	8	22	15
22	19	8	12	6	15	25	6	17	9
23	24	10	10	7	16	22	23	18	12
27	12	10	13	7	17	19	24	21	11
2	9	15	15	9	17	18	26	26	15
(0)	7	16	17	10	22	18	23	20	18
15	5	15	19	0	21	17	25	18	17

Table 1-2. Tally of Raw Data to Prepare Frequency Distribution.

Contamination Level (No. Brown Spots per Frequency	Group Frequency	Relative
0 – 5	7	0.07
6 – 11	23	0.23
12 – 17	27	0.27
18 – 23	31	0.31
24 – 29	10	0.10
30 – 35	2	0.02
	100	1.00

can now select groups and tally as shown in Table 1-2. The relative frequency column in Table 1-2 is calculated from the following definition.

$$\text{Relative Frequency} = \text{Group Frequency/Total Data} \qquad (1)$$

The frequency distribution can be represented graphically as shown in Figure 1-1. Figure 1-1a shows a line chart frequency distribution histogram, where the midpoint of each group set is used for the corresponding group frequency. Figure 1-1b shows a bar chart histogram.

On a personal computer using a spreadsheet program such as Lotus 1-2-3, this exercise should only take moments, once the data have been typed in.

MEASURES OF LOCATION, MEANS AND DATA CODING

The *measures of location* refer to the centrality or norm that characterizes a typical value of a population or sample population of data. The principle

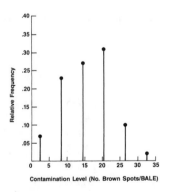

FIGURE 1-1a. Line chart histogram.

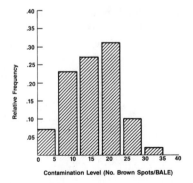

FIGURE 1-1b. Bar chart histogram.

term used to describe this property is the *mean*, of which there are several definitions (average, arithmetic mean, expected value, geometric mean). Symbols used are:

$$\mu = \text{population mean}$$

$$\bar{x} = \text{sample mean}$$

These are the most widely used measure of location and can best be related by analogy to the concept of center of gravity for bodies.

The mean value is computed from a data set $x_1, \ldots x_n$ as follows:

$$\bar{x} = \frac{\displaystyle\sum_{i=1}^{n} x_i}{n} \qquad (2)$$

The population mean, μ, is calculated on a similar basis.

Illustration 4

Calculate the mean value for the following set: {16, 12, 7, 11, 17, 9}:

$$\bar{x} = \frac{16 + 12 + 7 + 11 + 17 + 9}{6} = \frac{72}{6} = 12.0$$

The calculation of the mean of grouped data is more readily performed on the basis of a *weighted average*. Grouping the data will result in some loss of accuracy, compared with averaging ungrouped data.

Illustration 5

Tabulate the k group midpoints, x_i, and group frequencies f_i. Then compute the mean. Use the data set in Illustration 3.

From Table 1-2 we tabulate the following:

x_i	f_i	$f_i x_i$
2.5	7	17.5
8.5	23	195.5
14.5	27	391.5
20.5	31	635.5
26.5	10	265.0
32.5	2	65.0
	100	1570.0

The mean is calculated from:

$$\bar{x} = \frac{\sum\limits_{i=1}^{k} f_i x_i}{\sum\limits_{i=1}^{k} f_i} = \frac{1570}{100} = 15.70$$

Note that the mean value of the raw data was *15.47*.

The *median* denotes the 50*th* percentile value (i.e., the middlemost value). The symbol most often used is $\tilde{x}$. It is the second most important measure of location. It does, however, have limitations; it does not use the entire population of data and, hence, is less reliable than the mean as an estimate of μ. A further disadvantage is that it can be a cumbersome calculation for large samples. Consequently, it is most often used for small data samples. Whereas the mean cannot be used for ordered qualitative data, the median can.

The calculation procedure for the median is:

- First, order the data.
- For an odd-numbered sample size n: $\tilde{x} = x_{\left(\frac{n+1}{2}\right)}$

* For an even-numbered sample size n: $\tilde{x} = 1/2 \left(x_{\frac{n}{2}} + x_{\frac{n+2}{2}} \right)$

Illustration 6

Using the data set from Illustration 4:

$$x_{(1)} = 7, \quad x_{(2)} = 9, \quad x_{(3)} = 11, \quad x_{(4)} = 12, \quad x_{(5)} = 16, \quad x_{(6)} = 17$$

(a) For $n = 6$, $n/2 = 3$ and $(n + 2)/2 = 4$,

Therefore,

$$\tilde{x} = \frac{1}{2} (x_{(3)} + x_{(4)}) = \frac{1}{2} (11 + 12) = 11.5$$

(b) Assume $n = 5$; then $(n + 1)/2 = 3$

$$\tilde{x} = x_{(3)} = 11$$

The *mode* is a property value that corresponds to the highest frequency in the data. It can be thought of as the most likely value. If the data values are random, the mode may not exist. In addition, it may not be unique;

Table 1-3. Tabulations for Illustration 7.

Column No.	(I) x_i	(II) c	(III) z_i	(IV) f_i	(V) $f_i z_i$
	2.5	*14	*11.5	7	* 80.5
	8.5	*14	* 5.5	23	*126.5
	14.5	*14	0.5	27	13.5
	20.5	*14	6.5	31	201.5
	26.5	*14	12.5	10	125.0
	32.5	*14	18.5	2	37.0
					170.0

that is, two or more values may exist with equal frequency. It can be used for qualitative data having no natural ordering and in samples above the median point.

Manipulating or *coding* of data is a useful technique for simplifying statistical calculations. This is done by subtracting a number from each data value. The number chosen should be rounded off and approximately in the center of the data in order to obtain the coded data. The formula is:

$$z_i = x_i - c \tag{3}$$

Illustration 7

Using the data in Illustration 3, subtract a round number c approximately in the center of the set to obtain the coded data. Using the midpoint values x_i and from the frequency distribution in Table 1-2, we calculate the coded midpoint z_i (refer to columns (I) – (III) in Table 1-3).

Next, we compute the coded mean $\bar{z}$ *using the coded midpoints,* z_i

$$\bar{z} = \frac{\displaystyle\sum_{i=1}^{k} f_i z_i}{\displaystyle\sum_{i=1}^{k} f_i}$$

$f_i z_i$ and frequency in columns (IV) and (V) of Table 1-3. Therefore,

$$\bar{z} = \frac{170}{100} = 1.70$$

We now decode the value to obtain the true mean:

$$\bar{x} = \bar{z} + c$$
$$\bar{x} = 1.7 + 14.0 = 15.7$$

VARIABILITY OF DATA

The spread or dispersion of a data set provides a measure of the variability of the population about the mean. The parameter that describes this property is the *standard deviation* (also called the root mean square deviation). The symbols used are σ for the population standard deviation, and s for the sample standard deviation.

$$\sigma = \sqrt{\frac{\sum\limits_{i=1}^{N} (x_i - \mu)^2}{N}} \tag{4}$$

$$s = \sqrt{\frac{\sum\limits_{i=1}^{N} (x_i - \bar{x})^2}{n - 1}} \tag{5}$$

The units of σ and s are the same as the units of the data. The quantity $(n - 1)$ is referred to as the *degrees of freedom*. The deviations squared, σ^2 and s^2, are called the *variances* of the population and sample, respectively. An alternate formula for the sample standard deviation is:

$$s = \sqrt{\frac{\sum\limits_{i=1}^{n} x_i^2 - \dfrac{\left(\sum\limits_{i=1}^{n} x_i\right)^2}{n}}{n - 1}} \tag{6}$$

Illustration 8

For the data $x_1, \ldots x_n$ in Illustration 4, calculate the standard deviation s. From the data, $\bar{x} = 12.0$. The following values are tabulated:

x_i	$(x_i - \bar{x})$	$(x_i - \bar{x})^2$
16	4	16
12	0	0
7	−5	25
11	−1	1
17	5	25
9	−3	9
		76

By using formula (5), we get:

$$s = \sqrt{\frac{76}{(6-1)}} = 3.90$$

Or, with formula (6):

$$\sum_{i=1}^{n} x_i^2 = 256 + 144 + 49 + 121 + 289 + 81 = 940$$

$$\sum_{i=1}^{n} x_i = 72$$

$$s = \sqrt{\frac{940 - \frac{(72)^2}{6}}{5}} = \sqrt{\frac{940 - 864}{5}} = 3.90$$

The *sample range*, R, represents the difference between the largest and smallest value in the sample. It is most often used as a cross-check of the σ calculation. R does not use all the data and, hence, is not as accurate as s. Usefulness is limited to sample sizes less than 15.

Illustration 9

By using the data from Illustration 4, we obtain:

$$R = x_{(max)} - x_{(min)}$$

$$x_{(max)} = 17, x_{(min)} = 7$$

$$R = 17 - 7 = 10$$

The relative variation of data is called the *coefficient of variation*. It is defined as the ratio of the standard deviation to the mean and is expressed as a percentage:

$$r = \frac{\sigma}{\mu} \cdot 100 \tag{7}$$

Since it is a dimensionless quantity, it can be used to compare variability among different units of scale.

Illustration 10

Using the data from Illustration 8, calculate r for $\bar{x} = 12.0$, $s = 3.90$. Hence,

$$r = \frac{s}{\bar{x}} \cdot 100 = \frac{3.90}{12} \times 100 = \underline{3.25\%}$$

TYPES OF DISTRIBUTIONS

Previously we introduced distributions of discrete variates that take on a limited number of distinct values. Now we direct attention to variables that are continuous, such as height, weights, concentrations, or yields. The variable continues without a break from one value to the next with no limit to the number of distinct values. Consider the histogram of contamination levels in Figure 1-1b. Imagine that the size of the sample is increased without limit and the class intervals on the horizontal axis are decreased correspondingly. Figure 1-1b would gradually become a continuous curve. Continuous variables are distributed in a number of ways; we first consider the *normal distribution*.

The normal or Gaussian distribution is a symmetrical, bell-shaped curve, and is entirely determined by its mean μ and standard deviation σ. With a continuous variable x, we may define a function $f(x)$ that states that the height or ordinate of the continuous curve at the abscissa value x. The height $f(x)$ at the value x is:

$$Y = f(x) = [1/\sigma \sqrt{2\pi}]e^{-(x - \mu)^2/2\mu^2} \tag{8}$$

The theoretical Gaussian distribution curve is shown in Figure 1-2. The areas corresponding to the intervals shown are as follows:

Interval	Area (%)	Interval	Area (%)
$\mu \pm \sigma$	68.27	$\mu \pm 0.674\sigma$	50
$\mu \pm 2\sigma$	95.45	$\mu \pm 1.645\sigma$	90
$\mu \pm 3\sigma$	99.73	$\mu \pm 1.960\sigma$	95
$\mu \pm 4\sigma$	99.994	$\mu \pm 2.576\sigma$	99

We can interpret the Gaussian distribution in the following manner. If a very

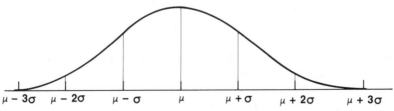

FIGURE 1-2. Gaussian distribution curve.

large sample were drawn from this population, we could expect 95.45% of the values to lie within the limits of $\mu \pm 2\sigma$.

The importance of the Gaussian distribution is as follows:

1 Many distributions of variables of natural phenomenon are approximately normal; examples include bubble sizes in fluidized reactors, natural soil particle sizes, pollutant concentrations in the atmosphere.

2 For measurements whose distributions are not normal, a simple transformation of the scale of measurements may approximate normality. These transformations often occur as a square root, $\sqrt{x}$, and the logarithm, ℓnx.

3 The normal distribution provides a simple mathematical relation, thus facilitating analytical solutions to many problems. This approach often provides reasonable approximations within engineering standards even when samples are derived from non-normal populations. This approach is often feasible even when handling populations with discrete variables.

4 Even when the distribution in the original population is highly non-normal, the distribution of the sample means $\bar{x}$ tends to approach normal under random sampling, as the sample size increases. This in itself is justification for the importance of the Gaussian distribution.

Since the normal curve depends on the two paramaters μ and σ, there are many different normal curves. Standard tables of this distribution given in statistics textbooks such as Snedecor and Cochran (1980), Walpole and Myers (1972), and Fisher (1950), are for the distribution with $\mu = 0$ and $\sigma = 1$. Therefore, for a measurement x with mean μ and standard deviation σ, in order to use a table of the normal distribution, one must rescale x so that the mean becomes 0 and the standard deviation becomes unity. The rescaled measurement can be calculated from:

$$z' = (x - \mu)/\sigma \tag{9}$$

Variable z' is called the *standard normal variate* (also the standard normal deviate or normal variate in standard measure). Values can be transformed back to the x-scale by:

$$x = \mu + \sigma z \tag{10}$$

A *skewed* distribution is one that is non-symmetrical. Skewness can be measured by the following formula:

$$\Sigma(x - \bar{x})^3/(\Sigma(x - \bar{x})^2)^{3/2} \tag{11}$$

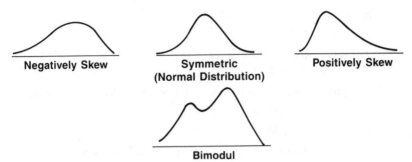

FIGURE 1-3. Different shapes of the distribution curve.

Figure 1-3 shows different shapes of the distribution curve.

A final note is in order on the distributions of sample means. If samples of size n are repeatedly taken from a population of mean μ and standard deviation σ, then the distribution of the averages $\bar{x}$ will have mean μ and a standard deviation $\sigma/\sqrt{n}$. When the parent distribution is Gaussian, then the distribution of averages will also be normal. This is illustrated in Figure 1-4 and below.

Illustration 11

Physical testing of several rubber samples at different periods of production shows an average tensile value of 1000 psi with a standard deviation of 25. If 6 lots could be blended, the distribution of the blends would have a mean tensile value of 1000 psi and a standard deviation of:

$$\frac{25}{\sqrt{6}} = 10.2$$

PRACTICE PROBLEMS

1 Compute the standard deviations of the following data sets as populations and as samples:

(a) 1, 5, 5, 6, 3, 2.

FIGURE 1-4. Shows the distribution of $\bar{x}$.

(b) 1, 2, 3, 3, 5, 6.

(c) 7, 7, 7, 5, 5, 5.

Answers: (a) $s = 1.97$, $\sigma = 1.80$; (b) $s = 1.86$, $\sigma = 1.70$;
(c) $s = 1.10$, $\sigma = 1.00$.

2 Compute the mean, median and mode of the following sets of data:

(a) 25, 10, 15, 18, 19, 27, 32, 5, 11, 6.

(b) 23, 13, 8, 3, 17, 7, 8.

Answers: (a) mean $\bar{x} = 16.8$, median $\tilde{x} = 16.5$, mode does not exist;
(b) mean $\bar{x} = 11.29$, median $\tilde{x} = 8$, mode is 8 and is not unique.

3 The heights of bales of rubber were measured as they came out of a compactor and loaded into crates for shipment. The heights in centimeters are given below. Prepare a frequency distribution, and compute the mean and standard deviation of the sample.

103	86	115	95	103	109
110	97	107	98	105	107
127	95	103	90	94	93
98	92	110	101	107	98
95	99	93	102	108	102

4 The specification for an injection-molded part requires that the pressure at a certain point not exceed 38 lb. A manufacturer who wants to compete in this market can make components with a mean pressure $\mu = 36.5$ lb, but the pressure varies among specimens with a standard deviation of 3.1 lb. Determine what percentage of the specimens will fail to meet the specification.

5 (a) Determine the mean and standard deviation of the population of values generated by the throw of a single die. (b) Repeat this for the throws of two dice.

Answers: (a) $\mu = 3.5$, $\sigma = 1.71$; (b) $\mu = 3.5$, $\sigma = 1.21$.

CHAPTER 2

Confidence Limits and Sample Size

CONFIDENCE INTERVAL

It is important when making an estimate of a population mean to have a measure of the accuracy of the estimate. For a random sample of size n from a population (n is large enough so that $\bar{x}$ can be assumed to be normally distributed), the probability that $\bar{x}$ is in error by less than plus or minus any given quantity can be calculated, provided σ is known for this population.

Illustration 1

In a manufacturing operation of sacks of cement, the distribution of weights per sack had $\sigma = 3.50$ kg. For a random sample of 400 sacks from this population, determine the probability that the sample mean is correct to within 100 grams.

The standard deviation of $\bar{x}$ is $\sigma/\sqrt{n} = 3500$ gms/20 = 175 gms.

The probability that the error $\bar{x} - \mu$ is less than 100 gms equals the probability that $z = (\bar{x} - \mu)/175$ is less than 100/175 = 0.571 in absolute value. Table 2-1 shows the cumulative normal frequency distribution (i.e., the area under the standard normal curve from 0 to z). From this table, the probability that z lies between 0 and 0.571 is (by interpolation) 0.216. Multiplying this by 2 to include the probability that z lies between -0.571 and 0, gives *0.43*.

Repeating this for the probability that $\bar{x}$ is correct to within 500 gms, the limit on $|z|$ is 2.86, and the probability desired is $2 \times 0.4979 = 0.996$, i.e., almost certainty.

The above illustration demonstrates the importance of the *confidence interval*. The confidence interval is a range of values which contains a population parameter with a certain degree of confidence. The *risk* taken in making this assertion is denoted by the symbol α, where $0 < \alpha < 1$. The *confidence level* is defined as $1 - \alpha$ [or expressed as a percentage, $100(1 - \alpha)$]. The end points in the confidence interval are referred to as the *confidence limits*.

A two-sided confidence interval is defined for the population mean μ, and is of the form $x \pm \delta$. That is, it is symmetrical around $\bar{x}$, where δ depends on the risk α, the estimate s and the sample size n. The procedure for applying the two-sided confidence interval is as follows:

- Select the desired confidence level, $1 - \alpha$.

Table 2-1. Cumulative Normal Frequency Distributions.

Z	0.00	0.01	0.02	0.03	0.04	0.05	0.06	0.07	0.08	0.09
0.0	0.0000	0.0040	0.0080	0.0120	0.0160	0.0199	0.0239	0.279	0.0319	0.0359
0.1	.0398	.0438	.0478	.0517	.0557	.0596	.0636	.0675	.0714	.0753
0.2	.0793	.0832	.0871	.0910	.0948	.0987	.1026	.1064	.1103	.1141
0.3	.1179	.1217	.1255	.1293	.1331	.1368	.1406	.1443	.1480	.1517
0.4	.1554	.1591	.1628	.1664	.1700	.1736	.1772	.1808	.1844	.1879
0.5	.1915	.1950	.1985	.2019	.2054	.2088	.2123	.2157	.2190	.2224
0.6	.2257	.2291	.2324	.2357	.2389	.2422	.2454	.2486	.2517	.2549
0.7	.2580	.2611	.2642	.2673	.2704	.2734	.2764	.2794	.2823	.2852
0.8	.2881	.2910	.2939	.2967	.2995	.3023	.3051	.3078	.3106	.3133
0.9	.3159	.3186	.3212	.3238	.3264	.3289	.3315	.3340	.3365	.3389
1.0	.3413	.3438	.3461	.3485	.3508	.3531	.3554	.3577	.3599	.3621
1.1	.3643	.3665	.3686	.3708	.3729	.3749	.3770	.3790	.3810	.3830
1.2	.3849	.3869	.3888	.3907	.3925	.3944	.3962	.3980	.3997	.4015
1.3	.4032	.4049	.4066	.4082	.4099	.4115	.4131	.4147	.4162	.4177
1.4	.4192	.4207	.4222	.4236	.4251	.4265	.4279	.4292	.4306	.4319
1.5	.4332	.4345	.4357	.4370	.4382	.4394	.4406	.4418	.4429	.4441
1.6	.4452	.4463	.4474	.4484	.4495	.4505	.4515	.4525	.4535	.4545
1.7	.4554	.4564	.4573	.4582	.4591	.4599	.4608	.4616	.4625	.4633
1.8	.4641	.4649	.4656	.4664	.4671	.4678	.4686	.4693	.4699	.4706
1.9	.4713	.4719	.4726	.4732	.4738	.4744	.4750	.4756	.4761	.4767
2.0	.4772	.4778	.4783	.4788	.4793	.4798	.4803	.4808	.4812	.4817
2.1	.4821	.4826	.4830	.4834	.4838	.4842	.4846	.4850	.4854	.4857
2.2	.4861	.4864	.4868	.4871	.4875	.4878	.4881	.4884	.4887	.4890
2.3	.4893	.4896	.4898	.4901	.4904	.4906	.4909	.4911	.4913	.4916
2.4	.4918	.4920	.4922	.4925	.4927	.4929	.4931	.4932	.4934	.4936
2.5	.4938	.4940	.4941	.4943	.4945	.4946	.4948	.4949	.4951	.4952
2.6	.4953	.4955	.4956	.4957	.4959	.4960	.4961	.4962	.4963	.4964
2.7	.4965	.4966	.4967	.4968	.4969	.4970	.4971	.4972	.4973	.4974
2.8	.4974	.4975	.4976	.4977	.4977	.4978	.4979	.4979	.4980	.4981
2.9	.4981	.4982	.4982	.4983	.4984	.4984	.4985	.4985	.4986	.4986
3.0	.4987	.4987	.4987	.4988	.4988	.4989	.4989	.4989	.4990	.4990
3.1	.4990	.4991	.4991	.4991	.4992	.4992	.4992	.4992	.4993	.4993
3.2	.4993	.4993	.4994	.4994	.4994	.4994	.4994	.4995	.4995	.4995
3.3	.4995	.4995	.4995	.4996	.4996	.4996	.4996	.4996	.4996	.4997
3.4	.4997	.4997	.4997	.4997	.4997	.4997	.4997	.4997	.4997	.4998
3.6	.4998	.4998	.4999	.4999	.4999	.4999	.4999	.4999	.4999	.4999
3.9	.5000									

- Calculate the arithmetic mean, $\bar{x}$, and the standard deviation, s (see Chapter 1).
- The two-sided Student's t chart is given in Table 2-2. Look up a value for t for $100(1 - \alpha)\%$ confidence and $n - 1$ degrees of freedom.

- Compute the following:

$$x_u = \bar{x} + ts/\sqrt{n}$$

$$x_L = \bar{x} - ts/\sqrt{n}$$

- This test enables us to conclude that we may assert with $100(1 - \alpha)\%$ confidence that:

$$x_L < \mu < x_u$$

Review the following illustration.

Table 2-2. Two-Sided Student's t Statistic.

Degrees of Freedom	100(1 − α)% Confidence Level		
	90%	95%	99%
1	6.314	12.706	63.657
2	2.920	4.303	9.925
3	2.353	3.182	5.841
4	2.132	2.776	4.604
5	2.015	2.571	4.032
6	1.943	2.447	3.707
7	1.895	2.365	3.499
8	1.860	2.306	3.355
9	1.833	2.262	3.250
10	1.812	2.228	3.169
11	1.796	2.201	3.106
12	1.782	2.179	3.055
13	1.771	2.160	3.012
14	1.761	2.145	2.977
15	1.753	2.131	2.947
16	1.746	2.120	2.921
17	1.740	2.110	2.898
18	1.734	2.101	2.878
19	1.729	2.093	2.861
20	1.725	2.083	2.845
21	1.721	2.080	2.831
22	1.717	2.074	2.819
23	1.714	2.069	2.807
24	1.711	2.064	2.797
25	1.708	2.060	2.787
26	1.706	2.056	2.779

(continued)

Table 2-2. (continued).

Degrees of Freedom	100(1 − α)% Confidence Level		
	90%	95%	99%
27	1.703	2.052	2.771
28	1.701	2.048	2.763
29	1.699	2.045	2.756
30	1.697	2.042	2.750
40	1.684	2.021	2.704
60	1.671	2.000	2.660
120	1.658	1.980	2.617
∞	1.645	1.960	2.576

Illustration 2

Determine the 90% confidence interval for reactor yields based on the following measurements {85.7, 89.6, 86.6, 93.2}.

- $1 - \alpha = 0.90$ or $\alpha = 0.10$
- $\bar{x} = 88.8$
- $s = 3.39$
- $t = 2.353$
- $x_u = 88.8 + (2.353) (3.39)/\sqrt{4} = 92.8$
  $x_L = 88.8 - (2.353) (3.39)/\sqrt{4} = 84.8$
- We may conclude that we are 90% confident that the true process yield lies between:

$$84.8 < \mu < 92.8$$

based on the four measurements.

This test is applied in cases where the criterion is that a minimum tolerance or value be met, or that a maximum value not be exceeded. In either case, the other end of the range is not a criterion. The procedure for applying this test is as follows:

- Select the desired confidence level, $1 - \alpha$.
- Calculate $\bar{x}$ and s.
- Obtain the t-value from Table 2-3 (values for one-sided Student's t statistics) for $100(1 - \alpha)$% confidence and $n-1$ degrees of freedom.
- Compute the following:

$$x_u = \bar{x} + ts/\sqrt{n} \text{ or } x_L = \bar{x} - ts/\sqrt{n},$$

depending on governing criterion.

- This test enables us to conclude that we may assert with $100(1 - \alpha)\%$ confidence that either:

$$\mu < x_u, \text{ or}$$

$$x_L < \mu$$

Illustration 3

Determine the upper 95% confidence bound for the process yields given in Illustration 2.

- $1 - \alpha = 0.95$ or $\alpha = 0.05$
- $\bar{x} = 88.8$
  $s = 3.39$
- $t = 2.353$
- $x_u = 88.8 + (2.353)(3.39)/\sqrt{4} = 92.8$
- We are 95% confident that the process yield cannot exceed 92.8% with the present operation.

SAMPLE SIZE AND ESTIMATING σ

In the above illustrations we have ignored an important consideration, namely, the size of the sample required to meet a specified accuracy of the conclusions. The investigator must state how accurate the sample should be. This can only be answered when a clear definition is given of the purposes to which the estimate will be put. Along with this, one must envision the consequences of having errors of different amounts in the estimate. If the estimate is to be made to guide an investment or commercialization decision regarding a new process, calculations may indicate the level of accuracy required to make an estimate useful. More often than not, however, this is the most difficult task in process development and applied research, and an element of arbitrariness is usually interjected into the final answer.

The accuracy or reliability of an estimate can be stated to be correct to within some limit $\pm \delta$. Recall that the normal distribution curve extends from $- \infty$ to $+ \infty$. Hence, we cannot guarantee that $\bar{x}$ is certain to lie between the limits $\mu - \delta$ and $\mu + \delta$. As shown above, however, we can make the probability that $\bar{x}$ lies between these limits as large as desired. In practice, this probability is usually established at 95% or 99%. At 95% probability, we know that there is a 95% chance that $\bar{x}$ lies between the limits of $\mu - 1.96/\sqrt{n}$ and $\mu + 1.96/\sqrt{n}$. In terms of the limit:

$$1.96\sigma/\sqrt{n} = \delta \tag{1}$$

This relation can be solved for n to obtain the required sample size. To use this relation, σ must be known, although the sample has not yet been drawn. One must therefore estimate σ, in which case historical data on the same or similar populations often provides a clue. Since this is an approximation, the coefficient 1.96 can be rounded off to a value of 2. Hence:

$$n = 4\sigma^2/\delta^2 \qquad (2)$$

For the 99% probability:

$$n = 6.6\sigma^2/\delta^2 \qquad (3)$$

In summary, given an estimate of σ and a selected risk α, we may wish to estimate the sample size required to determine μ within $\pm\delta$. This procedure provides a conservative estimate of the same size n, which is most realistic when σ is estimated using a large number of degrees of freedom. The procedure is as follows:

- Select δ and the allowable margin of error α, i.e., the risk that our estimate of μ is off by δ or more.
- Using Table 2-2, obtain a value of t for $100(1 - \alpha)$% confidence and ν degrees of freedom.
- Calculate the number of samples, n, required to estimate μ. This can be done using the following expression based on Table 2 use:

$$n = t^2 s^2/\delta^2 \qquad (4)$$

Round the value up to the nearest integer.
- We can conclude that if we calculate $\bar{x}$ from a sample size n, then we are $100(1 - \alpha)$% confident that:

$$\bar{x} - \delta < \mu < \bar{x} + \delta$$

Illustration 4

We wish to measure the residence time in an experimental reactor using a tracer technique. How many experiments must we run if we want to estimate the residence time at steady state to within 0.5%? We have an estimate of σ of 0.50 minutes with 3 degrees of freedom.

- We choose an α value of 5% risk and have decided $\delta = 0.5$ minutes.
- From Table 2-3, $t = 3.182$.
- Sample size is calculated

$$n = (3.182)^2 \, (0.50)^2/(0.5)^2$$

$$= 10.13 \cong 10$$

Table 2-3. One-Sided Student's t Statistic.

Degrees of Freedom	100(1 – α)% Confidence Level		
	90%	95%	99%
1	3.078	6.314	31.821
2	1.886	2.920	6.965
3	1.638	2.353	4.541
4	1.533	2.132	3.747
5	1.476	2.015	3.365
6	1.440	1.943	3.143
7	1.415	1.895	2.998
8	1.397	1.860	2.896
9	1.383	1.833	2.821
10	1.372	1.812	2.764
11	1.363	1.796	2.718
12	1.356	1.782	2.681
13	1.350	1.771	2.650
14	1.345	1.761	2.624
15	1.341	1.753	2.602
16	1.337	1.746	2.583
17	1.333	1.740	2.567
18	1.330	1.734	2.552
19	1.328	1.729	2.539
20	1.325	1.725	2.528
21	1.323	1.721	2.518
22	1.321	1.717	2.508
23	1.319	1.714	2.500
24	1.318	1.711	2.492
25	1.316	1.708	2.485
26	1.315	1.706	2.479
27	1.314	1.703	2.473
28	1.313	1.701	2.467
29	1.311	1.699	2.462
30	1.310	1.697	2.457
40	1.303	1.684	2.423
60	1.296	1.671	2.390
120	1.289	1.658	2.358
∞	1.282	1.645	2.326

- Hence, to be 95% sure of estimating the reactor's residence time to within 0.5%, we need a total of 10 experiments.

We now direct attention to the two-sided confidence interval for population standard deviation σ. The confidence interval in this case is not symmetrical around its estimate s. The reason for this is that large overestimates are usually made rather than underestimates. The two-sided interval technique is essentially used to show how much we really know about σ. For the researcher, this can be the governing criterion in establishing a new analytical method or tool. The following procedure can be used:

- Select the desired confidence level, $1 - \alpha$.
- Calculate the standard deviation s.
- Table 2-4 provides values of $(\sigma/s)_L$ for $100(1 - \alpha)\%$ confidence and $n - 1$ degrees of freedom.
- Compute the limits with:

$$s_u = s(\sigma/s)_u$$

$$s_L = s(\sigma/s)_L$$

(5)

- We conclude that we may assert with $100(1 - \alpha)\%$ confidence that:

$$s_u < \sigma < s_L$$

Table 2-4. Two-Sided σ/s Statistic.

| Degrees of Freedom | 100 $(1 - \alpha)\%$ Confidence Limits | | | | | |
| | 90% | | 95% | | 99% | |
	Lower	Upper	Lower	Upper	Lower	Upper
1	0.510	16.01	0.446	31.94	0.356	160.1
2	0.578	4.407	0.521	6.287	0.434	14.14
3	0.620	2.920	0.567	3.727	0.483	6.468
4	0.649	2.372	0.599	2.875	0.519	4.394
5	0.672	2.090	0.624	2.453	0.546	3.484
6	0.690	1.916	0.644	2.202	0.569	2.979
7	0.705	1.797	0.661	2.035	0.588	2.660
8	0.718	1.711	0.675	1.916	0.604	2.440
9	0.729	1.645	0.688	1.826	0.618	2.277
10	0.739	1.593	0.699	1.755	0.630	2.154
11	0.748	1.551	0.708	1.698	0.641	2.056
12	0.755	1.515	0.717	1.651	0.651	1.976
13	0.762	1.485	0.725	1.611	0.660	1.910
14	0.769	1.460	0.732	1.577	0.669	1.853
15	0.775	1.437	0.739	1.548	0.676	1.806
16	0.780	1.418	0.745	1.522	0.683	1.764
17	0.785	1.400	0.750	1.499	0.690	1.727
18	0.790	1.384	0.756	1.479	0.696	1.695
19	0.794	1.370	0.760	1.461	0.702	1.666
20	0.798	1.358	0.765	1.444	0.707	1.640

(continued)

Table 2-4. (continued).

| Degrees of Freedom | 100 (1 − α)% Confidence Limits | | | | | |
| | 90% | | 95% | | 99% | |
	Lower	Upper	Lower	Upper	Lower	Upper
22	0.805	1.335	0.773	1.415	0.717	1.595
24	0.812	1.316	0.781	1.391	0.726	1.558
26	0.818	1.300	0.788	1.370	0.734	1.526
28	0.823	1.286	0.794	1.352	0.741	1.499
30	0.828	1.274	0.799	1.337	0.748	1.475
35	0.838	1.248	0.811	1.304	0.762	1.427
40	0.847	1.228	0.821	1.280	0.774	1.390
45	0.854	1.212	0.829	1.259	0.784	1.361
50	0.861	1.199	0.837	1.243	0.793	1.337
55	0.866	1.188	0.843	1.229	0.801	1.316
60	0.871	1.179	0.849	1.217	0.808	1.299
65	0.875	1.170	0.854	1.207	0.814	1.285
70	0.879	1.163	0.858	1.198	0.820	1.272
80	0.886	1.151	0.866	1.183	0.829	1.250
90	0.892	1.141	0.873	1.171	0.838	1.233
100	0.897	1.133	0.879	1.161	0.845	1.219
120	0.905	1.120	0.888	1.145	0.856	1.196
140	0.911	1.110	0.895	1.133	0.866	1.179
160	0.916	1.102	0.901	1.123	0.873	1.166
180	0.921	1.096	0.906	1.115	0.880	1.155
200	0.925	1.090	0.911	1.109	0.885	1.146
300	0.937	1.072	0.926	1.087	0.904	1.116
400	0.945	1.062	0.935	1.074	0.916	1.099
500	0.951	1.055	0.942	1.066	0.924	1.088
600	0.955	1.050	0.946	1.060	0.930	1.080
800	0.961	1.043	0.953	1.052	0.939	1.068
1000	0.965	1.038	0.958	1.046	0.945	1.061
2000	0.975	1.027	0.970	1.032	0.961	1.042
5000	0.984	1.017	0.981	1.020	0.975	1.026
10000	0.989	1.012	0.986	1.014	0.982	1.019

Illustration 5

A simple laboratory technique is being developed to measure the friability of rubber bales. The test consists of measuring the stress relaxation on a small sample of rubber after a needle probe has been inserted at constant load. The time required to reach a certain level of stress relaxation can be related to baling conditions required in the plant to make a friable product. If 5 replicate measurements of the time required for relaxation to a single stress level are made on the same sample, what are the 95% confidence limits on σ? The data in seconds are 35, 37, 39, 36, 41.

- $1 - \alpha = 0.95$
- $s = 2.41$

- From Table 2-4 – $(\sigma/s)_u = 2.875$
$$(\sigma/s)_L = 0.599$$
- $s_u = 2.41(2.875) = 6.92$
$s_L = 2.41(0.599) = 1.44$
- We conclude that we are 95% confident that our standard deviation for the test is between:

$$1.44 < \sigma < 6.92 \text{ sec.}$$

A similar analysis can also be applied to a one-sided confidence interval for the population standard deviation. As noted earlier, a one-sided confidence interval for the σ is used when either a large value or minimum value is the limiting criterion for the analysis. An application example is the comparison of the variability of a new test method with an established technique that has a well-defined σ.

The procedure for the one-sided confidence interval test for the population standard deviation σ is as follows:

- Select the desired confidence level, $1 - \alpha$.
- Calculate the standard deviation, s.
- Use Table 2-5 to obtain a value of $(\sigma/s)_u$ or $(\sigma/s)_L$ for $100(1 - \alpha)\%$ confidence and $n - 1$ degrees of freedom.
- Calculate $s_u = s(\sigma/s)_u$ or $s_L = s(\sigma/s)_L$
- We conclude that we may assert with $100(1 - \alpha)\%$ confidence that:

$$\sigma < s_u \text{ or } s_L < \sigma$$

Table 2-5. One-Sided σ/s Statistic.

	Lower Confidence Limit				Upper Confidence Limit		
ν	90%	95%	99%	ν	90%	95%	99%
1	.607	.510	.388	1	7.906	16.01	79.06
2	.659	.578	.466	2	3.071	4.407	10.00
3	.693	.620	.514	3	2.264	2.920	5.110
4	.718	.649	.549	4	1.939	2.372	3.671
5	.735	.672	.576	5	1.762	2.090	3.004
6	.752	.690	.597	6	1.651	1.916	2.623
7	.762	.705	.616	7	1.571	1.797	2.377
8	.774	.718	.631	8	1.514	1.711	2.204
9	.783	.729	.645	9	1.470	1.645	2.076
10	.790	.739	.656	10	1.433	1.593	1.977
11	.798	.748	.667	11	1.404	1.551	1.898
12	.803	.755	.677	12	1.380	1.515	1.833

(continued)

Table 2-5. (continued).

	Lower Confidence Limit				Upper Confidence Limit		
ν	90%	95%	99%	ν	90%	95%	99%
13	.811	.762	.685	13	1.358	1.485	1.779
14	.816	.769	.693	14	1.341	1.460	1.733
15	.819	.775	.700	15	1.324	1.437	1.694
16	.825	.780	.707	16	1.311	1.418	1.659
17	.828	.785	.713	17	1.299	1.400	1.629
18	.833	.790	.719	18	1.287	1.384	1.602
19	.836	.794	.725	19	1.277	1.370	1.578
20	.839	.798	.730	20	1.268	1.358	1.556
22	.845	.805	.739	22	1.252	1.335	1.518
24	.851	.812	.747	24	1.238	1.316	1.487
26	.854	.818	.755	26	1.226	1.300	1.460
28	.860	.823	.762	28	1.216	1.286	1.437
30	.864	.828	.768	30	1.206	1.274	1.416
35	.870	.838	.781	35	1.188	1.248	1.375
40	.877	.847	.792	40	1.174	1.228	1.343
45	.884	.854	.802	45	1.162	1.212	1.318
50	.891	.861	.810	50	1.152	1.199	1.297
55	.894	.866	.818	55	1.143	1.188	1.280
60	.898	.871	.824	60	1.137	1.179	1.265
65	—	.875	.830	65	—	1.170	1.252
70	.905	.879	.835	70	1.125	1.163	1.241
80	.909	.886	.844	80	1.116	1.151	1.222
90	.913	.892	.852	90	1.108	1.141	1.207
100	.921	.897	.858	100	1.102	1.133	1.195
120	.925	.905	.869	120	1.092	1.120	1.175
140	.928	.911	.877	140	1.085	1.110	1.160
160	.933	.916	.884	160	1.078	1.102	1.148
180	.937	.921	.890	180	1.073	1.096	1.139
200	.941	.925	.895	200	1.070	1.090	1.131
300	.949	.937	.913	300	1.056	1.072	1.104
400	.958	.945	.924	400	1.048	1.062	1.089
500	.962	.951	.931	500	1.042	1.055	1.079
600	—	.955	.937	600	—	1.050	1.072
800	—	.961	.945	800	—	1.043	1.062
1000	.971	.965	.950	1000	1.030	1.038	1.055
2000	—	.975	.964	2000	—	1.027	1.038
5000	.990	.984	.977	5000	1.013	1.017	1.024
10000	—	.989	.984	10000	—	1.012	1.017

Illustration 6

Determine the 95% confidence upper bound on the standard deviation in the previous illustration.

- $1 - \alpha = 0.95$
- $s = 2.41$
- $(\sigma/s)_u = 2.372$

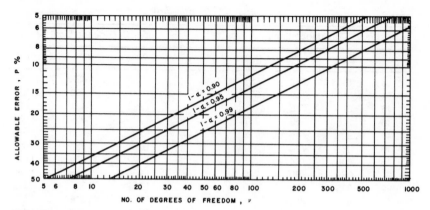

FIGURE 2-1. Chart for estimating the number of measurements required to estimate σ within $P\%$ of its true value.

- $s_u = (2.41)(2.372) = 5.72$
- We are 95% confident that the standard deviation of the test is less than 5.72 seconds.

A plot that can be used to estimate the number of measurements required to estimate σ within $P\%$ of its true value is given in Figure 2-1. To use this chart, select a value for P for an allowable percent error α. That is, we establish the risk that our estimate of σ is off by $P\%$ or more. From the values of P and α, Figure 2-1 can be used to obtain the number of degrees of freedom. From the degrees of freedom, ν, the number of measurements needed to estimate σ can be calculated with this formula:

$$n = \nu + 1$$

Illustration 7

Determine the number of measurements required to estimate σ within 10% of its true value.

- $P = 10\%$, and we'll choose $1 - \alpha = 0.95$ ($\alpha = 0.05$).
- From Figure 2-1, $\nu = 188$
- $n = \nu + 1 = 189$

We would have to make 189 repeat experiments to be 95% confident that s was within 10% of σ. Obviously we have selected too stringent a requirement.

PROBABILITY CHART PAPER

Many statistical methods impose the assumption of normality. It is therefore necessary to check the normality of a data set before selecting the method of analysis to apply. A graphic check can be made using arithmetic probability paper, an example of which is shown in Figure 2-2. The abscissa scale is linear; however, the ordinates are in units of percentages, descending on the left scale from 99.99 to 0.01 and rising on the right scale from 0.01 to 99.99. The ordinate scale is stretched at both ends—i.e., the distance from 50 to 60 is smaller than that from 60 to 70, which is smaller than that from 70 to 80, etc. This distortion is provided so that if we plot for any x the percentage of the standard normal population that exceeds x, the points lie on a straight line. The straight line represents the standard normal population. The *s*-shaped

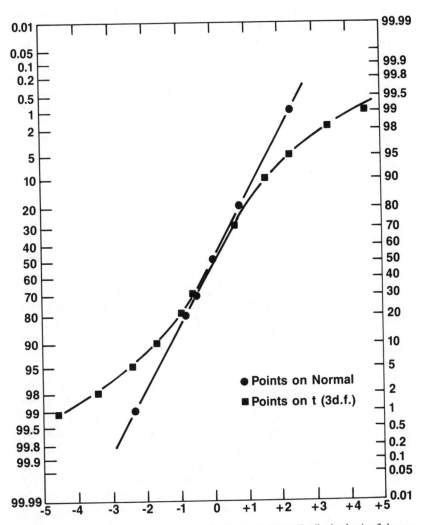

FIGURE 2-2. Plots of the cumulative normal distribution and the *t*-distribution having 3 degrees of freedom, on probability paper.

curve is the corresponding plot for a t distribution having 3 degrees of freedom. Although this latter distribution is symmetrical, it has much longer tails than the normal.

When checking the normality of a large sample of data, one should plot for each of a set of values of x the percentage of the sample that exceeds x. An alternate approach is to plot for each x the percentage of the sample that is less than x, using the right-hand ordinate scale. The purpose of these plots is to look for systematic deviations of the type which are indicative of a long-tailed distribution, or those indicating a skewed distribution.

With small samples, it may be desirable to plot the probability corresponding to the lowest value of x. For this value, the sample percentage less than x is zero, which of course cannot be plotted on probability paper. Snedecor and Cochran (1980) show an easy way to avoid this by arguing that the lowest value of a sample size of 10 for example, ought to lie approximately in the middle of the lowest tenth of the frequency distribution. On this basis then it should be plotted at 5% probability less than x. At the same time, the second lowest x should be plotted at 15%, and so forth. This rule plots the ith x from the bottom in a sample of size n versus $100(2i - 1)/(2n)\%$ less than x. The references given at the end of this book describe graphical and numerical tests for skewness and kurtosis (i.e., presence of long- or short-tails), which are only outlined in this manual.

TESTS OF HYPOTHESES

Statistical analysis relies heavily on the so-called *test of a hypothesis* (or the *test of significance*). The importance of this test is that it can play the deciding role in decision-making or analysis. For example, let's say we want to assess two competing but different technologies of comparable capital investments for the manufacture of phthalic anhydride, and that one of the deciding factors is the overall process yield. Before reporting the average differences found in the data, we would be well advised to ask: Do the results show real differences among the average conversions in different populations, or can the differences observed in the experiments be explained by process variations from run to run? Thus, we might set up a simple hypothesis that $\mu_A = \mu_B$ (where A and B refer to the two different technologies). A typical analysis might have several hypotheses which should be examined to ensure that sample results support each hypothesis. Unless our analyses support these hypotheses (or reject them depending on criteria established), we are in no position to claim that there are real differences among populations in the comparison of different technologies, analytical procedures, etc. In statistics, the hypothesis under test is referred to as the *null-hypothesis*. In this subsection, the mechanics of the principle tests used are illustrated. The user should refer to the Bibliographies at the end of the book for detailed explanation of the theory and for further illustrations.

A null-hypothesis test is essentially a formal decision-making process employing statistical criteria. The procedure can be summarized in a general manner by a sequence of steps:

Step 1 The hypothesis to be tested must be stated as a single criterion or series of mathematical criteria. In other words, a population parameter is set equal to a specified value, or a range of values.

Step 2 Select the risk α of rejecting the hypothesis when it is actually true.

Step 3 Select an appropriate test statistic that can be computed from the experimental data sheet.

Step 4 Determine the critical region of the test statistic which will reject the hypothesis, based on the α-value.

Step 5 Compute the test statistic.
a) If the value falls in the critical region, the hypothesis is rejected with risk α.
b) If the statistic falls outside the critical region then the hypothesis is not rejected.

Step 6 In the event that the hypothesis is not rejected, another type of error arises, namely, that of accepting the hypothesis when it is actually false. In this situation, the parameter is actually equal to a value outside the hypothesis region. The risk of this occurrence is denoted by the symbol β, and is dependent on the difference between the actual and the hypothesized values. It is also strongly dependent upon the sample size used.

Statistical hypothesis tests may be two-sided or one-sided depending on the analysis applied. We first consider the two-sided hypothesis test with application to the population mean μ. The objective in this problem class is to assess whether or not the population mean is different from some bench mark value (say μ_o), with respect to measuring error. The hypothesis is stated in the following manner:

$$H_o : \mu = \mu_o, \text{ or}$$

$$H_o : \mu - \mu_o = 0$$

where the symbol H_o denotes that this is a null-hypothesis test. The reason why this is referred to as a two-sided hypothesis test is that we are concerned with values of μ both greater than and less than μ_o. The following steps outline the procedure for this test:

- Choose the risk α of asserting a difference when none exists.
- From Table 2-2, obtain a value of t for $100(1 - \alpha)\%$ confidence and $n - 1$ degrees of freedom.

- Calculate the arithmetic mean $\bar{x}$ and the standard deviation s.
- Use the following formula to compute the statistical parameter:

$$t^* = |\bar{x} - \mu_o| / (s/\sqrt{n}) \tag{6}$$

- We may arrive at one of the following conclusions:
  - If $t^* \geq t$, then we may assert that $\mu \neq \mu_o$ with $100(1 - \alpha)\%$ confidence.
  - If $t^* < t$, then we can conclude that there is no basis to believe that μ differs from the standard deviation, μ_o.

Illustration 8

A product specification in the manufacture of terpolymers is the slope of the curemeter data when the material is compounded in an ASTM compound formulation. In a particular production run the following curemeter performances were measured on different samples: 23, 25, 22, 19, 21 in-lb/min. Can we release the product for sale with a 95% confidence that we met the sales spec of 20 in-lb/min curemeter performance?

- $\alpha = 0.05$
- From Table 2-2, $t = 2.776$
- $\bar{x} = 22.0$
  $s = 2.24$
- $t^* = |22 - 20| / (2.24/\sqrt{5}) = 2.00$
- Since $t^* < t$ there is not enough evidence to support that the product is out-of-spec and should not be shipped.

We now direct attention to the one-sided hypothesis test, again with application to the population mean μ. In this problem class the interest is only that either $\mu > \mu_o$ or $\mu < \mu_o$. The direction of the hypothesis is a critical element and depends on the objectives of the analysis. The hypothesis can be chosen so that it represents the greater risk if the hypothesis is falsely rejected. This risk becomes controlled at the α level. The procedure in this test is as follows:

- Select the risk α of asserting that a difference exists when there is no difference; i.e.,

$$H_o : \mu > \mu_o$$

$$H_o : \mu < \mu_o$$

- Use Table 2-3 to obtain t for $100(1 - \alpha)\%$ confidence and $n - 1$ degrees of freedom.
- Calculate $\bar{x}$ and s.

- Calculate the statistical parameter

$$t^* = (\mu - \bar{x})/(s/\sqrt{n}) \qquad (7A)$$

$$t^* = (\bar{x} - \mu_o)/(s/\sqrt{n}) \qquad (7B)$$

- We may arrive at one of the following conclusions:
  - If $t^* > t$, we may assert that $\mu < \mu_o$ or that $\mu > \mu_o$ with $100(1 - \alpha)\%$ confidence.
  - If $t^* < t$, we conclude that there is no basis to believe that $\mu < \mu_o$ (or that $\mu > \mu_o$).

Illustration 9

Two elastomer products are blended for a special customer that requires an 87 Shore A hardness in the final blend. Measurements of several batches are 88.1, 86.9, 89.6, 88.1 and 87.5 Shore A. Can we state with 90% confidence that the blends supplied will be better than the customer's minimum requirement?

- $\alpha = 0.10$
- $t = 1.533$
- $\bar{x} = 88.0$
  $s = 1.00$
- $t^* = (\bar{x} - \mu_o)/(s/\sqrt{n})$
  $= (88.0 - 87)/(1.00/\sqrt{s})$
  $= 2.24$
- Since $t^* > t$, we may conclude with 90% confidence that our blends more than meet the customer's minimum hardness requirement.

In applying a null-hypothesis test it is important to note that the sample size should be sufficiently large to insure that differences of economic importance will be noted with a reasonably high probability. At the same time, the sample size should not be so large that unimportant differences become magnified. The following outlines the steps that should be applied in the determination of the proper sample size in the hypothesis tests on μ. The objective here is to determine how many measurements are needed in order to compare a mean to a standard value so that we are able to detect a difference ($= \mu - \mu_o$) with a risk β of missing the difference.

- Select the α-risk (i.e., the risk of asserting that there is a difference when none exists).
- Select β, the risk of failing to detect a difference when one of δ exists.
- Decide whether a two-sided or one-sided test should be made after the data are assembled.

Table 2-6. Number of Observations for t-Test of Mean.

Value of $\delta = \dfrac{\mu - \mu_0}{\sigma}$	Single-sided test $\alpha = 0.005$ Double-sided test $\alpha = 0.01$					$\alpha = 0.01$ $\alpha = 0.02$					$\alpha = 0.025$ $\alpha = 0.05$					$\alpha = 0.05$ $\alpha = 0.1$				
$\beta =$	0.01	0.05	0.1	0.2	0.5	0.01	0.05	0.1	0.2	0.5	0.01	0.05	0.1	0.2	0.5	0.01	0.05	0.1	0.2	0.5
0.05																				
0.10																				
0.15																				122
0.20										139					99					70
0.25					110									128	64			139	101	45
0.30				134	78				115	63			119	90	45		122	97	71	32
0.35			125	99	58			109	85	47		109	88	67	34		90	72	52	24
0.40		115	97	77	45		101	85	66	37	117	84	68	51	26	101	70	55	40	19
0.45		92	77	62	37	110	81	68	53	30	93	67	54	41	21	80	55	44	33	15
0.50	100	75	63	51	30	90	66	55	43	25	76	54	44	34	18	65	45	36	27	13
0.55	83	63	53	42	26	75	55	46	36	21	63	45	37	28	15	54	38	30	22	11
0.60	71	53	45	36	22	63	47	39	31	18	53	38	32	24	13	46	32	26	19	9
0.65	61	46	39	31	20	55	41	34	27	16	46	33	27	21	12	39	28	22	17	8
0.70	53	40	34	28	17	47	35	30	24	14	40	29	24	19	10	34	24	19	15	8
0.75	47	36	30	25	16	42	31	27	21	13	35	26	21	16	9	30	21	17	13	7
0.80	41	32	27	22	14	37	28	24	19	12	31	22	19	15	9	27	19	15	12	7
0.85	37	29	24	20	13	33	25	21	17	11	28	21	17	13	8	24	17	14	11	6
0.90	34	26	22	18	12	29	23	19	16	10	25	19	16	12	7	21	15	13	10	6
0.95	31	24	20	17	11	27	21	18	14	9	23	17	14	11	7	19	14	11	9	5
1.00	28	22	19	16	10	25	19	16	13	9	21	16	13	10	6	18	13	11	8	5

(continued)

Table 2-6. (continued).

	Level of t-test																			
Single-sided test	$\alpha = 0.005$					$\alpha = 0.01$					$\alpha = 0.025$					$\alpha = 0.05$				
Double-sided test	$\alpha = 0.01$					$\alpha = 0.02$					$\alpha = 0.05$					$\alpha = 0.1$				
$\beta =$	0.01	0.05	0.1	0.2	0.5	0.01	0.05	0.1	0.2	0.5	0.01	0.05	0.1	0.2	0.5	0.01	0.05	0.1	0.2	0.5
1.1	24	19	16	14	9	21	16	14	12	8	18	13	11	9	6	15	11	9	7	
1.2	21	16	14	12	8	18	14	12	10	7	15	12	10	8	5	13	10	8	6	
1.3	18	15	13	11	8	16	13	11	9	6	14	10	9	7		11	8	7	6	
1.4	16	13	12	10	7	14	11	10	9	6	12	9	8	7		10	8	7	5	
1.5	15	12	11	9	7	13	10	9	8	6	11	8	7	6		9	7	6		
1.6	13	11	10	8	6	12	10	9	7	5	10	8	7	6		8	6	6		
1.7	12	10	9	8	6	11	9	8	7		9	7	6	5		8	6	5		
1.8	12	10	9	8	6	10	8	7	7		8	7	6			7	6			
1.9	11	9	8	7	6	10	8	7	6		8	6	6			7	5			
2.0	10	8	8	7	5	9	7	7	6		7	6	5			6				
2.1	10	8	7	7		8	7	6	6		7	6				6				
2.2	9	8	7	6		8	7	6	5		7	6				6				
2.3	9	7	7	6		8	6	6			6	5				5				
2.4	8	7	7	6		7	6	6			6									
2.5	8	7	6	6		7	6	6			6									
3.0	7	6	6	5		6	5	5			5									
3.5	6	5	5			5														
4.0	6																			

Value of $\delta = \dfrac{\mu - \mu_0}{\sigma}$

31

Table 2-7. Number of Observations for t-Test of Difference Between Two Means.

Single-sided test → Double-sided test → $\beta =$	α = 0.005 / α = 0.01					α = 0.01 / α = 0.02					α = 0.025 / α = 0.05					α = 0.05 / α = 0.1				
Value of $\delta = \dfrac{\mu_1 - \mu_1}{\sigma}$	0.01	0.05	0.1	0.2	0.5	0.01	0.05	0.1	0.2	0.5	0.01	0.05	0.1	0.2	0.5	0.01	0.05	0.1	0.2	0.5
0.05																				
0.10																				
0.15																				
0.20																				137
0.25															124					88
0.30										123					87					61
0.35					110					90					64				102	45
0.40					85					70				100	50			108	78	35
0.45				118	68				101	55			105	79	39		108	86	62	28
0.50				96	55			106	82	45		106	86	64	32		88	70	51	23
0.55			101	79	46		106	88	68	38		87	71	53	27	112	73	58	42	19
0.60		101	85	67	39		90	74	58	32	104	74	60	45	23	89	61	49	36	16
0.65		87	73	57	34	104	77	64	49	27	88	63	51	39	20	76	52	42	30	14
0.70	100	75	63	50	29	90	66	55	43	24	76	55	44	34	17	66	45	36	26	12
0.75	88	66	55	44	26	79	58	48	38	21	67	48	39	29	15	57	40	32	23	11
0.80	77	58	49	39	23	70	51	43	33	19	59	42	34	26	14	50	35	28	21	10
0.85	69	51	43	35	21	62	46	38	30	17	52	37	31	23	12	45	31	25	18	9
0.90	62	46	39	31	19	55	41	34	27	15	47	34	27	21	11	40	28	22	16	8
0.95	55	42	35	28	17	50	37	31	24	14	42	30	25	19	10	36	25	20	15	7
1.00	50	38	32	26	15	45	33	28	22	13	38	27	23	17	9	33	23	18	14	7

Level of t-test

(continued)

Table 2-7. (continued).

Level of t-test

| Single-sided test | α = 0.005 | | | | | α = 0.01 | | | | | α = 0.025 | | | | | α = 0.05 | | | | |
| Double-sided test | α = 0.01 | | | | | α = 0.02 | | | | | α = 0.05 | | | | | α = 0.1 | | | | |
β =	0.01	0.05	0.1	0.2	0.5	0.01	0.05	0.1	0.2	0.5	0.01	0.05	0.1	0.2	0.5	0.01	0.05	0.1	0.2	0.5
1.1	42	32	27	22	13	38	28	23	19	11	32	23	19	14	8	27	19	15	12	6
1.2	36	27	23	18	11	32	24	20	16	9	27	20	16	12	7	23	16	13	10	5
1.3	31	23	20	16	10	28	21	17	14	8	23	17	14	11	6	20	14	11	9	5
1.4	27	20	17	14	9	24	18	15	12	8	20	15	12	10	6	17	12	10	8	4
1.5	24	18	15	13	8	21	16	14	11	7	18	13	11	9	5	15	11	9	7	4
1.6	21	16	14	11	7	19	14	12	10	6	16	12	10	8	5	14	10	8	6	4
1.7	19	15	13	10	7	17	13	11	9	6	14	11	9	7	4	12	9	7	6	3
1.8	17	13	11	10	6	15	12	10	8	5	13	10	8	6	4	11	8	7	5	
1.9	16	12	11	9	6	14	11	9	8	5	12	9	7	6	4	10	7	6	5	
2.0	14	11	10	8	6	13	10	9	7	5	11	8	7	6	4	9	7	6	4	
2.1	13	10	9	8	5	12	9	8	7	5	10	8	6	5	3	8	6	5	4	
2.2	12	10	8	7	5	11	9	7	6	4	9	7	6	5		8	6	5	4	
2.3	11	9	8	7	5	10	8	7	6	4	9	7	6	5		7	5	5	4	
2.4	11	9	8	6	5	10	8	7	6	4	8	6	5	4		7	5	4	4	
2.5	10	8	7	6	4	9	7	6	5	4	8	6	5	4		6	5	4	3	
3.0	8	6	6	5	4	7	6	5	4	3	6	5	4	4		5	4	3		
3.5	6	5	5	4	3	6	5	4	4		5	4	4	3		4	3			
4.0	6	5	4	4		5	4	4	3		4	4	3			4				

Value of $\delta = \dfrac{\mu_1 - \mu_0}{\sigma}$

33

- Obtain an estimate of σ. Historical data for the experiment help to establish the σ-value.
- Table 2-6 provides the number of observations required for the t-test of the mean. Use this table to obtain the number of experiments required, n, based on the appropriate α.

Illustration 10

How many measurements are needed to detect a 2% increase in moisture 95% of the time in a drying operation? The standard deviation is approximately 0.5%.

- We select $\alpha = 0.01$
- Choose $\beta = 0.05$
- We decide to make a one-sided comparison once the data have been collected.
- σ has been estimated to be 0.5% from historical drying measurements in the plant.
- $\delta/\sigma = 2.0/0.5 = 4.0$

From Table 2-6, $n = 8$. Hence, we need to make 8 measurements in order to confirm that moisture levels have crept up, of course maintaining the risk of error as low as desired.

Table 2-7 provides the number of observations for the t-test applied to the difference between two means. The procedure to handle this problem class is similar to the above analysis.

OUTLIERS AND TOLERANCE INTERVALS

As is often the case in experimental programs, there is always a handful of data ponts that don't seem to agree with the trends of the population samples. This leads us to the subject of *outliers*. An outlier is a sample value that seemingly does not belong with the sample data, and often our first response is to reject it. It may in fact be a legitimate data point that reflects the effect of an independent variable on the process or phenomenon under investigation. A statistical test can be applied to ascertain the credibility of this suspected value. The reader, however, should be cautioned that the rejection of the outlier on purely statistical grounds is still risky. One should always try to define the basis for rejection by physical means, such as faulty instrumentation, inaccurate metering, poor operator attention, etc. At any rate, it's good practice to report what fraction of the population has been rejected based on statistical evaluations along with possible operating causes which substantiate their rejection. The following statistical procedure can be applied in establishing the credibility of outliers.

- Tabulate the n observation in either ascending or descending order, beginning with the suspected outlier.

Table 2-8. Critical Values of Range Test for Outliers.

Statistic	Number of obs., n	Critical values						
		$\alpha = .30$	$\alpha = .20$	$\alpha = .10$	$\alpha = .05$	$\alpha = .02$	$\alpha = .01$	$\alpha = .005$
$r_{10} = \dfrac{X_{(2)} - X_{(1)}}{X_{(n)} - X_{(1)}}$	3	.684	.781	.886	.941	.976	.988	.994
	4	.471	.560	.679	.765	.846	.889	.926
	5	.373	.451	.557	.642	.729	.780	.821
	6	.318	.386	.482	.560	.644	.698	.740
	7	.281	.344	.434	.507	.586	.637	.680
$r_{11} = \dfrac{X_{(2)} - X_{(1)}}{X_{(n-1)} - X_{(1)}}$	8	.318	.385	.479	.554	.631	.683	.725
	9	.288	.352	.441	.512	.587	.635	.677
	10	.265	.325	.409	.477	.551	.597	.639
$r_{21} = \dfrac{X_{(3)} - X_{(1)}}{X_{(n-1)} - X_{(1)}}$	11	.391	.442	.517	.576	.638	.679	.713
	12	.370	.419	.490	.546	.605	.642	.675
	13	.351	.399	.467	.521	.578	.615	.649
$r_{22} = \dfrac{X_{(3)} - X_{(1)}}{X_{(n-2)} - X_{(1)}}$	14	.370	.421	.492	.546	.602	.641	.674
	15	.353	.402	.472	.525	.579	.616	.647
	16	.338	.386	.454	.507	.559	.595	.624
	17	.325	.373	.438	.490	.542	.577	.605
	18	.314	.361	.424	.475	.527	.561	.589
	19	.304	.350	.412	.462	.514	.547	.575
	20	.295	.340	.401	.450	.502	.535	.562
	21	.287	.331	.391	.440	.491	.524	.551
	22	.280	.323	.382	.430	.481	.514	.541
	23	.274	.316	.374	.421	.472	.505	.532
	24	.288	.310	.367	.413	.464	.497	.524
	25	.262	.304	.380	.406	.457	.489	.516

- Choose the α-risk value, i.e., the risk of rejecting the data point, $X_{(1)}$, when it is not an outlier.
- Use Table 2-8 to calculate the critical values of the range test for outliers. The formulas to use are:

$$r_{10}^* = (x_{(2)} - x_{(1)})/(x_{(n)} - x_{(1)})$$

$$r_{11}^* = (x_{(2)} - x_{(1)})/(x_{(n-1)} - x_{(1)})$$

$$r_{21}^* = (x_{(3)} - x_{(1)})/(x_{(n-1)} - x_{(1)})$$ (8)

$$r_{22}^* = (x_{(3)} - x_{(1)})/(x_{(n-2)} - x_{(1)})$$

- Obtain the critical range value (r_{10} or r_{11} or r_{21} or r_{22}) from Table 2-8 for the appropriate sample size, n, and risk, α.
- Apply the following criteria to determine if the data value is an outlier:
  - If $r_{10}^* > r_{10}$, reject x as an outlier.
  - If $r_{10}^* < r_{10}$, there is insufficient evidence to reject $x_{(1)}$ as an outlier. Hence, $x_{(1)}$ must be retained for further analysis.

Illustration 11

The mean bubble size in a fluidized bed reactor was measured over a period of time during steady-state operation. The values obtained were 10.5 cm, 11.0 cm, 9.8 cm and 5.5 cm. Is the value of 5.5 cm an outlier?

- $x_{(1)} = 5.5$
  $x_{(2)} = 9.8$
  $x_{(3)} = 10.5$
  $x_{(4)} = 11.0$
- We are willing to take a 10% risk of being wrong if $x_{(1)}$ is not an outlier.
  $\alpha = 0.10$.
- $r_{10}^* = (9.8 - 5.5)/(11.0 - 5.5)$
  $= 0.782$
- From Table 2-8, $r_{10} = 0.679$
- $r_{10}^* > r_{10}$, hence, we may reject $x_{(1)}$ as being an outlier at the 90% confidence level.

The final subject for this chapter concerns *tolerance intervals*. A tolerance interval is an interval which will cover a fixed proportion of a population, P, with a specified confidence level $(1 - \alpha)$. The method assumes the distribution to be normal. The following procedure can be used:

- Select a desired confidence level $(1 - \alpha)$.

Table 2-9. Values of K for Tolerance Intervals of a Normal Distribution.

n	1 − α = .75					1 − α = .90					1 − α = .95					1 − α = .99				
P	0.750	0.900	0.950	0.990	0.999	0.750	0.900	0.950	0.990	0.999	0.750	0.900	0.950	0.990	0.999	0.750	0.900	0.950	0.990	0.999
2	4.498	6.301	7.414	9.351	11.920	11.407	15.978	18.800	24.167	30.227	22.858	32.019	37.674	48.430	60.573	114.363	160.193	188.491	242.300	303.054
4	2.035	2.992	3.431	4.471	5.657	2.932	4.166	4.943	6.440	8.149	3.779	5.369	6.370	8.299	10.502	6.614	9.398	11.150	14.527	18.383
6	1.704	2.429	2.889	3.779	4.802	2.196	3.131	3.723	4.870	6.188	2.604	3.712	4.414	5.775	7.337	3.743	5.337	6.345	8.301	10.548
8	1.568	2.238	2.663	3.491	4.444	1.921	2.743	3.264	4.278	5.446	2.197	3.136	3.732	4.891	6.226	2.905	4.147	4.936	6.468	8.234
10	1.492	2.131	2.537	3.328	4.241	1.775	2.535	3.018	3.959	5.064	1.987	2.839	3.379	4.433	5.649	2.508	3.582	4.265	5.594	7.129
12	1.443	2.062	2.456	3.223	4.110	1.683	2.404	2.863	3.758	4.792	1.858	2.655	3.162	4.150	5.291	2.274	3.250	3.870	5.079	6.477
14	1.409	2.013	2.398	3.148	4.016	1.619	2.314	2.736	3.618	4.619	1.770	2.529	3.012	3.955	5.045	2.120	3.029	3.608	4.737	6.043
16	1.383	1.977	2.355	3.092	3.946	1.572	2.246	2.676	3.514	4.484	1.705	2.437	2.903	3.812	4.865	2.009	2.872	3.421	4.492	5.732
18	1.363	1.948	2.321	3.048	3.891	1.535	2.194	2.614	3.433	4.382	1.655	2.366	2.819	3.702	4.725	1.926	2.753	3.279	4.307	5.497
20	1.347	1.925	2.294	3.013	3.840	1.506	2.152	2.564	3.368	4.300	1.616	2.310	2.752	3.615	4.614	1.860	2.659	3.168	4.161	5.312
22	1.334	1.906	2.271	2.984	3.809	1.482	2.118	2.524	3.315	4.232	1.584	2.264	2.697	3.543	4.523	1.808	2.584	3.078	4.044	5.163
24	1.322	1.891	2.252	2.959	3.778	1.462	2.089	2.489	3.270	4.176	1.557	2.225	2.651	3.483	4.447	1.764	2.522	3.004	3.947	5.039
30	1.297	1.853	2.210	2.904	3.708	1.417	2.025	2.413	3.170	4.049	1.497	2.140	2.549	3.350	4.278	1.668	2.385	2.841	3.733	4.768
35	1.283	1.834	2.183	2.871	3.667	1.390	1.988	2.368	3.112	3.974	1.462	2.090	2.490	3.272	4.179	1.613	2.306	2.748	3.611	4.611
40	1.271	1.818	2.166	2.846	3.635	1.370	1.959	2.334	3.066	3.917	1.435	2.053	2.445	3.213	4.104	1.571	2.247	2.677	3.518	4.493
45	1.262	1.805	2.150	2.826	3.609	1.354	1.935	2.306	3.030	3.871	1.414	2.021	2.408	3.165	4.042	1.539	2.200	2.622	3.444	4.399
50	1.255	1.794	2.138	2.809	3.588	1.340	1.916	2.284	3.001	3.833	1.396	1.996	2.379	3.126	3.993	1.512	2.162	2.576	3.385	4.323

(continued)

Table 2-9. (continued).

P	1 − α = .75					1 − α = .90					1 − α = .95					1 − α = .99				
	0.750	0.900	0.950	0.990	0.999	0.750	0.900	0.950	0.990	0.999	0.750	0.900	0.950	0.990	0.999	0.750	0.900	0.950	0.990	0.999
60	1.243	1.778	2.118	2.784	3.556	1.320	1.887	2.248	2.955	3.774	1.369	1.958	2.333	3.066	3.916	1.471	2.103	2.506	3.293	4.206
70	1.235	1.765	2.104	2.764	3.531	1.304	1.865	2.222	2.920	3.730	1.349	1.929	2.299	3.021	3.859	1.440	2.060	2.454	3.225	4.120
80	1.228	1.756	2.092	2.749	3.512	1.292	1.848	2.202	2.894	3.696	1.334	1.907	2.272	2.986	3.814	1.417	2.026	2.414	3.173	4.053
90	1.223	1.748	2.083	2.737	3.497	1.283	1.834	2.185	2.872	3.669	1.321	1.889	2.251	2.958	3.778	1.398	1.999	2.382	3.130	3.999
100	1.218	1.742	2.075	2.727	3.484	1.275	1.822	2.172	2.854	3.646	1.311	1.874	2.233	2.934	3.748	1.383	1.977	2.355	3.096	3.954
120	1.211	1.732	2.063	2.712	3.464	1.262	1.804	2.150	2.826	3.610	1.304	1.850	2.205	2.898	3.702	1.358	1.942	2.314	3.041	3.885
140	1.206	1.724	2.054	2.700	3.449	1.252	1.791	2.134	2.804	3.582	1.282	1.833	2.184	2.870	3.666	1.340	1.916	2.283	3.000	3.833
160	1.202	1.718	2.047	2.691	3.437	1.245	1.780	2.121	2.787	3.561	1.272	1.819	2.167	2.848	3.638	1.326	1.896	2.259	2.968	3.792
180	1.198	1.713	2.042	2.683	3.427	1.239	1.771	2.111	2.774	3.548	1.264	1.808	2.154	2.831	3.616	1.314	1.879	2.239	2.942	3.759
200	1.195	1.709	2.037	2.677	3.419	1.234	1.764	2.102	2.762	3.529	1.258	1.798	2.143	2.816	3.597	1.304	1.865	2.222	2.921	3.731
250	1.190	1.702	2.028	2.665	3.404	1.225	1.750	2.085	2.740	3.501	1.245	1.780	2.121	2.788	3.561	1.286	1.839	2.191	2.880	3.678
300	1.186	1.695	2.021	2.656	3.393	1.217	1.740	2.073	2.725	3.481	1.236	1.767	2.106	2.767	3.535	1.273	1.820	2.169	2.850	3.641
400	1.181	1.688	2.012	2.644	3.378	1.207	1.726	2.057	2.703	3.453	1.223	1.749	2.084	2.739	3.499	1.255	1.794	2.138	2.809	3.589
500	1.177	1.683	2.008	2.636	3.368	1.201	1.717	2.046	2.689	3.434	1.215	1.737	2.070	2.721	3.475	1.243	1.777	2.117	2.783	3.555
600	1.175	1.680	2.002	2.631	3.360	1.196	1.710	2.038	2.678	3.421	1.209	1.729	2.060	2.707	3.458	1.234	1.764	2.102	2.763	3.530
700	1.173	1.677	1.998	2.626	3.355	1.192	1.705	2.032	2.670	3.411	1.204	1.722	2.052	2.697	3.445	1.227	1.747	2.091	2.748	3.511
800	1.171	1.675	1.996	2.623	3.350	1.189	1.701	2.027	2.663	3.402	1.201	1.717	2.046	2.688	3.434	1.222	1.741	2.082	2.736	3.495
900	1.170	1.673	1.993	2.620	3.347	1.187	1.697	2.023	2.658	3.396	1.198	1.712	2.040	2.682	3.426	1.218	1.736	2.075	2.726	3.483
1000	1.169	1.671	1.992	2.617	3.344	1.185	1.695	2.019	2.634	3.390	1.195	1.709	2.036	2.676	3.418	1.214	1.735	2.068	2.718	3.472
∞	1.130	1.645	1.960	2.576	3.291	1.150	1.645	1.960	2.576	3.291	1.150	1.645	1.960	2.576	3.291	1.150	1.645	1.960	2.576	3.291

- Choose the population proportion P that the interval is to cover.
- Compute $\bar{x}$ and s for the sample size n.
- Refer to standard statistics textbooks such as those at the end of this chapter, for values of coefficient K for the tolerance intervals of a normal distribution. Table 2-9 provides a sample of such a chart. From this table, obtain K for $100(1 - \alpha)\%$ confidence, the proportion P and the sample size.
- Use the following formulas:

$$x_u = \bar{x} + Ks \qquad (9A)$$

$$x_L = \bar{x} - Ks \qquad (9B)$$

- From the above, we may assert with $100(1 - \alpha)\%$ confidence that 100 $P\%$ of the population lies between x_L and x_u.

Illustration 12

Determine the 95% tolerance interval for 75% of future production capacity based on yearly average plant production rates for the last 5 years (89, 92.1, 88.7, 95.3, 100%).

- $1 - \alpha = 0.95$ or 95% confidence
- $P = 0.75$ (i.e., 75% of the population is to be included)
- $\bar{x} = 93.0$
  $s = 4.73$
- From Table 2-9, $K = 3.779$
- $x_u = 93 + (3.779)(4.73) = 110.9$
  $x_L = 93 - (3.779)(4.73) = 76.3$
- We are 95% confident that 75% of future production capacity will lie between 76.3 and 110.9%, based on the last 5 years of production capacity. This would suggest that a plant expansion study might be in order.

PRACTICE PROBLEMS

1 Show that if the limit of error is reduced from δ to $\delta/2$, the sample size must be increased by a factor of 4. With the same δ, if we desire 99% probability of being within the limit rather than 95% probability, what percent increase in sample size is needed?

Answer: + 73%.

2 The heights of a random sample of rubber bales (sample size 16) from a population with $\sigma = 26$ cm are measured. What is the probability that $\bar{x}$ does not differ from μ by more than 10 cm.

Answer: Probability = 0.876.

3 In the sale of off-spec wax-oils, the manufacturer attempts to estimate the average price he can get for his off-spec products to within $\pm\$20$ per ton, apart from a 1-in-20 chance, based on the selling prices of his competitors. If the manufacturer guesses that σ is about \$60 based on previous years, how many competitor prices should be included in the sample?

Answer: $n = 36$.

4 Measurements of the die swell of six samples of an experimental polymer were as follows: 20%, 18%, 25%, 15%, 12%, 22%. What can we conclude about the true die swell of this polymer with a 90% confidence level?

5 Drying improvements have been made to the finishing operation in the production of our elastomers. The volatiles measurements made on several spot samples were 50 ppm, 70 ppm, 35 ppm, 15 ppm, 23 ppm and 29 ppm. What is the new variability of our volatiles spec on finished product expressed as a 90% confidence interval on σ?

6 Five repeat measurements of antioxidant levels in a rubber sample were made; test values are 0.151 wt. %, 0.158, 0.150, 0.152, 0.155. The sample has a known amount of antioxidant of 0.150%. Do we have a bias in our measurement technique?

7 Determine the number of samples we would have to test to determine if a shipment of catalyst feedstock is less than the purity specification by 0.2%. The test method has a standard deviation of 0.25%. Management is willing only to run a 0.5% risk of missing such a difference, and a 1% risk of being wrong when we claim that the shipment is out of specification to the manufacturer.

8 Determine whether the following observations are normally distributed, or can the high value of 10.2 be rejected as an outlier with 90% confidence?

5.3	4.8	3.7	10.2
6.1	5.1	3.9	4.9
1.1	6.0	4.2	5.2
1.3	4.9	6.5	4.5

CHAPTER 3

Data Scatter and the Use of Control Charts

CONTROL CHARTS

Control charts are used for detecting variations in the mean and for assessing the variability of a parameter or phenomenon in a time-monitored process. The charts are constructed from small data samples that are continually recorded. This data is used to estimate the process mean and variability, both over discrete time periods and cumulatively, and are subsequently compared with pre-established *control limits*. This chronological assessment of the process can be used on a real time basis to ascertain whether or not the process is in a *state of statistical control*. Figure 3-1 conceptually illustrates that a discrete sampling population of a particular variable can change over the course of the process operation. On a real time basis then, independent operating variables can be changed to bring the sample population back into the range of the control limits.

Although the use of control charts and statistical process control is most efficiently handled on computers, the mechanics of the analysis are illustrated below. The general strategy in constructing and applying control charts is as follows:

- Choose a subgroup size n. Subdivide the data into small groups, or sample from the process at a fixed frequency.
- Calculate the mean $\bar{X}_i$ and the sample range R_i for each group.
- Prepare plots of $\bar{X}_i$ and R_i versus the subgroup sequence number.
- Compute the overall average $\bar{\bar{X}}$ *and the average range* $\bar{R}$.

$$\bar{\bar{X}} = \frac{\sum\limits_{i=1}^{k} \bar{X}_i}{k} \tag{1}$$

$$\bar{R} = \frac{\sum\limits_{i=1}^{k} R_i}{k} \tag{2}$$

41

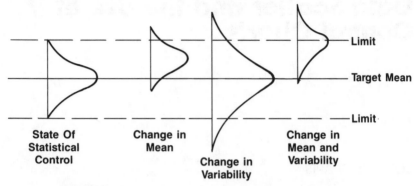

FIGURE 3-1. Illustrates statistical control of a process via its distribution.

where k = number of subgroups.

The calculation should be performed on the basis of a minimum of 10 to 25 subgroups. The values should be drawn as solid lines on the $\bar{X}$ and R plots.

- Next, compute 3σ control limits for averages and ranges in the following manner:

	$\bar{X}$ Chart	R Chart
Upper Control Limit (UCL)	$\bar{\bar{X}} + A_2\bar{R}$	$D_4\bar{R}$
Lower Control Limit (LCL)	$\bar{\bar{X}} - A_2\bar{R}$	$D_3\bar{R}$

The coefficients A_2, D_3, D_4 are given in Table 3-1.

The control limits should be indicated on the charts as dotted lines. These limits will be symmetric about $\bar{\bar{X}}$, but not about $\bar{R}$.

The charts can be interpreted in the following manner. First, a point outside the control limits on the R-chart indicates a shift in the process variability. Secondly, for the $\bar{X}$-chart, the following conditions indicate that the process mean has shifted: a point outside the control limits; two points in succession in the $2\sigma - 3\sigma$ region; and runs of points on the same side of X: 7 out of 7, 10 out of 11, 12 out of 14, 14 out of 17, 16 out of 20.

To maintain and upgrade control charts, the following guidelines should be followed. First, control limits should be systematically updated. This should be done at a minimum frequency of every 50–100 subgroups. Second, it is important to define a realistic criterion in order to satisfy that the process is in a state of statistical control. For practical purposes we may state that the process is controlled if no more than 0 out of 25, 1 out of 35, or 2 out of 100 data ponts are outside the limits. A reference point to use is the *process capability* which is the interval 6σ, where σ can be approximated as $\bar{R}/d_2$.

An important consideration in the chart development is the size of the subgroup. As a rule of thumb, a subgroup size of 4 or 5 will usually ensure a normally distributed mean. Smaller subgroup sizes can result in non-normal $\bar{X}$s. However, both time required for measurements and costs often influence the subgroup size. It is possible to employ subgroups of size unity for so-called *individual control charts*. In this situation $\bar{R}$ is estimated as the average of the moving ranges (i.e., successive differences). A final note is that the larger subgroup sizes selected, the more sensitive they are to small shifts in the process mean ($\pm\sigma$). When the subgroup size > 15, it is recommended that a σ-chart can be employed in lieu of an R-chart.

In the illustration that follows, we shall sequentially construct the control charts for several product quality parameters. In this example, we are concerned with the production of a copolymer in which the process operation is controlled to meet requirements of on-spec product. The product parameters to control are the viscosity of the polymer (Mooney viscosity) and the ethylene content of the polymer.

Table 3-1. Coefficients for Control Charts.

Subgroup Size	Control Limit Factors			Estimation of σ^*
	Average Chart	Range Chart		
n	A_2	D_3	D_4	d_2
1**	2.660	0	3.267	1.128
2	1.880	0	3.267	1.128
3	1.023	0	2.575	1.693
4	0.729	0	2.282	2.059
5	0.577	0	2.115	2.326
6	0.483	0	2.004	2.534
7	0.419	0.076	1.924	2.704
8	0.373	0.136	1.864	2.847
9	0.337	0.184	1.816	2.970
10	0.308	0.223	1.777	3.078
11	0.285	0.256	1.774	3.173
12	0.266	0.284	1.716	3.258
13	0.249	0.308	1.692	3.336
14	0.235	0.329	1.671	3.407
15	0.223	0.348	1.652	3.472
Control Limits	Averages	Ranges		
Upper	$\bar{\bar{X}} + A_2\bar{R}$	$D_4\bar{R}$		
Lower	$\bar{\bar{X}} - A_2\bar{R}$	$D_3\bar{R}$		

*Estimate of $\sigma = \bar{R}/d_2$
**Individual Control Charts: use moving ranges to estimate R.

Table 3-2. Data for Illustration 1.

Sample No.	Mooney Viscosity Mean	Mooney Viscosity Range	Ethylene (wt. %) Mean	Ethylene (wt. %) Range
1	78.5	2	58	1
2	76	2	76.9	1.5
3	77	3	80.6	0.7
4	86	5	71.7	0.6
5	73.5	6	72.3	0.3
6	76	4	73.5	0.2
7	83	4	73.2	0.4
8	103	5	73	0.6
9	74	5	73.5	0.5
10	72	2	73.4	0.6
11	72	3	73.5	0.4
12	83	4	73.2	0.2
13	80	7	73.4	0.5
14	81	5	70.1	0.1
15	79	4	73.1	0.1
16	79.5	4	72.8	0.5
17	82	3	72.9	0.4
18	86	2	73.2	0.3
19	93.5	2	73.4	0.2
20	85.5	4	74.4	0.2
21	74	3	75.4	0.2
22	83	5	74.5	0.3
23	90	5	74.5	0.4
24	84	6	76.3	0.6
25	93	6	77.3	0.4
26	100	2	79	0.4
27	102	2	74.3	0.5
28	84	3	71.4	0.5
29	81	2	72.5	0.5
30	74	3	73.5	0.7
31	83.5	2	73.6	0.8
32	64	4	73.6	0.6
33	73	5	73.3	0.7
34	73	3	73.7	0.7
35	75.5	2	74.3	0.3
36	67	3	72.2	0.2
37	73	5	71.1	0.2
38	71	5	70.2	0.4
39	85	5	70.5	0.5
40	80	4	71.8	0.5
41	102	4	70.3	0.4
42	85	2	70.9	0.3
43	72	3	71.1	0.2
44	76.5	2	72	0.1

Illustration 1

Mooney viscosity and ethylene content are obtained from polymer reactor samples at different times during the initial plant production of a market development grade. The measurements are obtained in triplicate and mean values are reported in Table 3-2. The specs for the product were established by the product development group based on property performance. In the case of acceptable Mooney product, the spec called for 75–85 Mooney product (i.e., a target product Mooney of 80). The product ethylene target called for a target mean of 73 wt. % with control to within 1 wt. %.

For Mooney control we first calculate the grand average $\bar{\bar{X}}$ and the average range $\bar{R}$:

$$\bar{\bar{X}} = 3566/44 = 81.1$$
$$\bar{R} = 162/44 = 3.7$$

From Table 3-1, for $n = 3$:

$$A_2 = 1.023, \quad D_3 = 0, \quad D_4 = 2.575$$

For the range chart:

$$UCL = D_4\bar{R} = 2.575 \times 3.7 = 9.48$$
$$LCL = D_3\bar{R} = 0$$

For the average chart:

$$UCL = \bar{\bar{X}} + A_2\bar{R} = 81.1 + (1.023)(3.7) = 84.9$$
$$LCL = \bar{\bar{X}} - A_2\bar{R} = 81.1 - 3.79 = 77.31$$

Figure 3-2a shows the control chart for the average Mooney. As shown, process operations chose 3σ control limits for the averages and actually targeted for a product that was slightly above the mean. Simply examining Figure 3-2 we would conclude that the process was not in a state of statistical control for a large portion of the plant trial.

The range chart for the Mooney is shown in Figure 3-2b. In comparing the plots in Figure 3-2 we can conclude that the variability of the product during this run did not change, but the process mean underwent dramatic swings.

We now direct attention to the ethylene content.

$$\bar{\bar{X}} = 3213.4/44 = 73.03$$
$$\bar{R} = 19.7/44 = 0.45$$

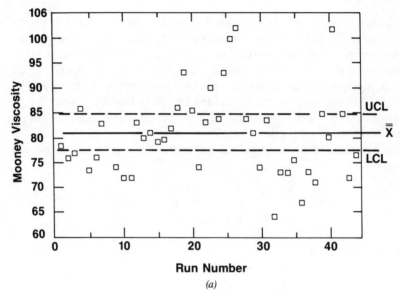

FIGURE 3-2a. Plot of mean viscosity data.

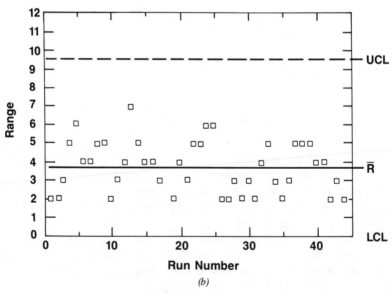

FIGURE 3-2b. Plot of range.

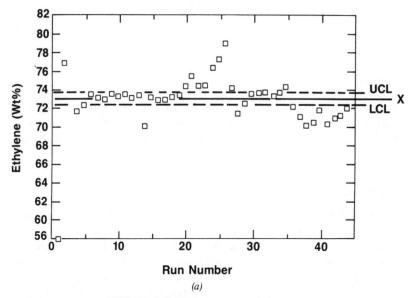

FIGURE 3-3a. Plot of mean ethylene data.

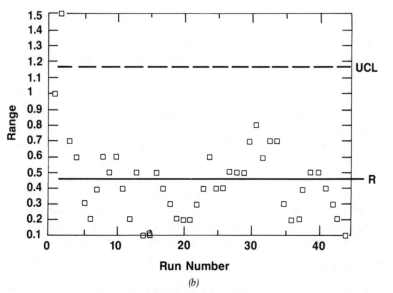

FIGURE 3-3b. Plot of range.

Range: UCL $= 2.575 \times 0.45 = 1.16$
        LCL $= 0$

Average:
        UCL $= 73.03 + 1.023 \times 0.45 = 73.5$
        LCL $= 73.03 - 0.46 = 72.6$

The average and range control charts are shown in Figure 3-3. Again, we may conclude that the process variability was quite stable, but a significant fraction of the population average was outside the control limits. As shown, better control of the ethylene spec was achieved than in the case of Mooney control.

CUMULATIVE SUM CONTROL CHARTS

Statistical process control is further facilitated with the use of Cumulative Sum Control Charts (Cusum charts), which plot the cumulative sum of deviations from the target value against time. This type of plot has the advantage of being more sensitive than the standard control charts illustrated above to small shifts in the process mean. They are not as frequently used as the standard control charts because they are more difficult to prepare and sometimes require a greater degree of interpretation.

The procedure for constructing and using Cusum charts is as follows:

- Specify a target value for the process. In the illustration given above the appropriate target could be the mean.
- For each X_i data, compute the deviation from the target value (i.e., calculate $X_i - \bar{X}$) and the cumulative sum (i.e., Cusum) from $Y_i = Y_{i-1} + (X_i - \bar{X})$.
- Prepare a plot of the Cusums versus the order of the data (or time).

The finished chart lends itself to the following interpretations. First, if the Cusums fluctuate in a horizontal fashion, then the process is likely on target. If the Cusums tend to deviate from the horizontal (i.e., from the target) at a constant angle, then a shift in the process mean has occurred. A simple statistical test to detect a change in the process mean involves the placement of a V-mask over the last point. If a point falls outside the V, a change has occurred at that point. The articles by Johnson and Leone (1962) and Lucas (1976) give detailed discussions on the use of this type of control chart. The following example continues with the analysis of Illustration 1.

Illustration 2

For Mooney viscosity control in Illustration 1, the target mean is 81 Mooney points. Table 3-3 tabulates the values for constructing the cumulative sum control chart. Figure 3-4 shows the chart. For the first 18 runs, the cumulative sums are shown to fluctuate about the horizontal segment A-B, indicating that

Table 3-3. Tabulations for Mooney Control for Illustration 2.

No.	Xi	Xi-target	Yi
1	78.5	−2.5	−2.5
2	76	−5	−7.5
3	77	−4	−11.5
4	86	5	−6.5
5	73.5	−7.5	−14
6	76	−5	−19
7	83	2	−17
8	103	22	5
9	74	−7	−2
10	72	−9	−11
11	72	−9	−20
12	83	2	−18
13	80	−1	−19
14	81	0	−19
15	79	−2	−21
16	79.5	−1.5	−22.5
17	82	1	−21.5
18	86	5	−16.5
19	93.5	12.5	−4
20	85.5	4.5	0.5
21	74	−7	−6.5
22	83	2	−4.5
23	90	9	4.5
24	84	3	7.5
25	93	12	19.5
26	100	19	38.5
27	102	21	59.5
28	84	3	62.5
29	81	0	62.5
30	74	−7	55.5
31	83.5	2.5	58
32	64	−17	41
33	73	−8	33
34	73	−8	25
35	75.5	−5.5	19.5
36	67	−14	5.5
37	73	−8	−2.5
38	71	−10	−12.5
39	85	4	−8.5
40	80	−1	−9.5
41	102	21	11.5
42	85	−4	15.5
43	72	−9	6.5
44	76.5	−4.5	2

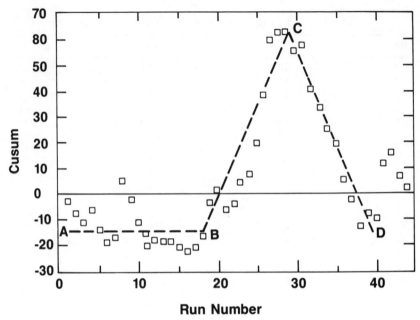

FIGURE 3-4. Cumulative sum control chart.

the process was reasonably well controlled. However, between runs 19–37, the Cusums deviate from the horizontal at a constant angle, indicating that the mean Mooney product being made had shifted. Corrective action was taken during the production run to bring the mean back on target as shown by the decreasing line segment C-D.

INTRODUCTION TO SPECTRAL ANALYSIS

Spectral analysis of time series data is applied in the frequency domain. The most common manner of evaluating time-dependent processes is in terms of an amplitude versus time function, known as a *time history record* of events. The instantaneous amplitude of the record represents the specific process parameter of interest. The technique has perhaps been most extensively applied in research activities such as the study of flow properties and disturbances [see Cheremisinoff (1986)]. A time scale of the record can represent some appropriate independent variable. In the case of fluid dynamic studies, the independent variable may be relative time, spacial location, or angular position. The advantage of these types of records is that for some physical phenomena it is possible to predict the specific time history records of future measurements with reasonable confidence based on an understanding of the physics and/or developing appropriate scale factors from prior experimental observations. For example, one can predict the mean drop size from a spray nozzle, the size of rising bubbles in an extraction column, the amount of particulate entrainment from a fluidized bed dryer, the efficiency of a cyclone separator, and the yields or the conversions of feedstocks in a chemical reactor. In these examples, the ability to predict certain parameters with confidence depends on how good a mathematical description of the phenomena is obtained and the range over which extrapolations are valid. Phenomena allowing prediction are referred to as *deterministic*. It should be noted that the phenomena of interest may not be deterministic. That is, each experiment results in a unique time history record that is non-repetitive. These types of data and physical disturbances are referred to as being random. The general interpretations and applications of random data analysis are introduced in this subsection. The basis of random data analysis is the use of correlation and spectral density functions, which require a working knowledge of the fundamentals of Fourier series and transform analysis as well as the basic response properties of physical systems. As an introduction to random data analysis some of the characteristics of random data are described below.

The resulting time history record from an experiment revealing random fluctuations represents only one physical realization of the events taking place in the flow system. To understand the entire time history of events a collection of a series of records must be analyzed. This is referred to as an *ensemble* and it defines a random process of the phenomenon. An ensemble is illustrated in Figure 3-5, where the collection of all time history records is denoted by $x_i(t)$, $i = 1,2,3 \ldots$, and the random process function is $\{x(t)\}$.

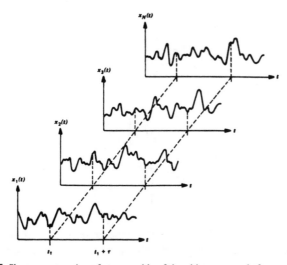

FIGURE 3-5. Shows construction of an ensemble of time history records for a random process.

Once an ensemble of time history events is obtained, average properties of the data can be derived for any specific time t_1 simply by averaging over the ensemble. Two very common averages are the *mean value* and *mean square value* which are the average value and the average squared value at time t_1, respectively. Mathematically these are expressed as:

$$\mu_x(t_1) = \lim_{N \to \infty} \frac{1}{N} \sum_{i=1}^{N} x_i(t_1) \tag{3}$$

$$\psi_x^2(t_1) = \lim_{N \to \infty} \frac{1}{N} \sum_{i=1}^{N} x_i^2(t_1) \tag{4}$$

The average product of the data value at times t_1 and $t_1 + \tau$ (where τ is a displacement or time delay) is the autocorrelation.

$$R_{xx}(t_1, \tau) = \lim_{t \to \infty} \frac{1}{N} \sum_{i=1}^{N} x_i(t_1) \, x_i \, (t_1 + \tau) \tag{5}$$

In the most general case, where average values of interest vary with time, data is referred to as being nonstationary. A limiting case is where all average values of interest remain constant with changes in time; data are referred to as *stationary* in this case. The average values at all times can be calculated from the appropriate ensemble averages at a single time t_1 for stationary data.

The average values in the above equations can be computed in most cases as follows for stationary values:

$$\mu_x = \lim_{T \to \infty} \frac{1}{T} \int_0^T x(t)dt \tag{6}$$

$$\psi_x^2 = \lim_{T \to \infty} \frac{1}{T} \int_0^T x^2(t)dt \tag{7}$$

$$R_{xx}(\tau) = \lim_{T \to \infty} \int_0^T x(t)x(t + \tau)dt \tag{8}$$

The basis for Equations (6) through (8) is the *ergodic theorem* which states that for stationary data, the properties computed from time averages over individual records of the ensemble are the same from one record to another and are equal to corresponding properties computed from an ensemble average over the records at any time t, provided that:

$$\frac{1}{T} \int_{-T}^{T} |R_{xx}(\tau) - \mu_x^2| \, d\tau \rightarrow 0; \ \tau \rightarrow \infty \tag{9}$$

The criterion of Equation (9) does not hold when periodic functions emerge in the data. Equation (9) is a sufficient condition for ergodicity, meaning that time averages can be justified even when periodic properties exist.

In practice, the limiting operations as $N \rightarrow \infty$ or $T \rightarrow \infty$ are not necessary. In other words, the number of time history records that are needed for analysis by ensemble averaging procedures, or the length of a given sample record required for analysis by time averaging methods, can be a finite value. The value of these will depend on the reproducibility of the average values of the parameters of interest. The error associated with the computation of an average value due to finite sampling conditions is of major importance to the interpretations and applications of processed data.

Effort should always be given to designing the experiment such that stationary data are produced. The analysis procedures for nonstationary data are much more complex, often requiring tremendously larger volumes of data to obtain meaningful statistical properties. In many experiments the results can be forced to a stationary state by maintaining certain conditions of the experiment constant. For example, if one is interested in studying the interfacial wave structure of stratified gas-liquid flow inside a pipe, stationary data will be generated if the liquid velocity and physical properties of the two fluids are held constant during each experiment (i.e., only the gas velocity is varied during the experiment).

There are of course, a class of exceptions where the basic parameters of the mechanisms generating the data are acts of nature that cannot be controlled by the experimenter. Examples are time history data for dispersion in the atmosphere and ocean, and the formation of by-products in a chemical reaction. In the former cases, it is not possible to design repetitive experiments to generate a meaningful ensemble. The general approach in analyzing such data is to select segments of the ensemble that are sufficiently long to provide statistically meaningful results. There are non-stationary models which permit the data to be regressed into stationary random and non-stationary deterministic components of analysis; however, this topic is beyond this short volume. Bendat and Piersol (1980) and Walpole and Myers (1972) provide introductory readings on non-stationary models.

In the most general sense, if there are N possible equally likely and mutually exclusive outcomes of an event (for example, tossing a coin) to a random or chance experiment, and if N_A of these outcomes correspond to an event A

of interest, then the probability of event A, $P(A)$ is:

$$P(A) = \frac{N_A}{N} \tag{10}$$

In defining relative frequency, we consider a random experiment, enumerate all possible outcomes, repeatedly perform the experiment, and take the ratio of outcomes, N_A, favoring the event of interest, A, to the total number of trials, N. This provides an approximation of the probability of A, $P(A)$. The limit of N_A/N is referred to as the relative frequency of A, as $N \rightarrow \infty$ as $P(A)$:

$$P(A) \overset{\Delta}{=} \underset{N \rightarrow \infty}{\text{Lim}} \frac{N_A}{N} \tag{11}$$

The relative frequency notation of probability is useful for estimating a probability. In engineering analyses, probability is best defined in terms of the relative frequency of occurrences. Another way of stating Equation (11) is that the probability of outcome A is the fractional portion of experimental results with the attribute A in the limit as the experiment is repeated for an infinite number of times.

Consider an ensemble of measurements where interest is in a particular measurement at some time t_1, that is, ϕ units or less. In other words, $A = x(t_1) < \phi$. The probability of occurrence is:

$$P[x(t_1) \leq \phi] = \underset{N \rightarrow \infty}{\text{lim}} \frac{N[x(t_1) \leq \phi]}{N} \tag{12}$$

$N[x(t_1) \leq \phi]$ is the number of measurements having amplitudes less than or equal to ϕ at time t_1. If amplitude ϕ is assigned arbitrary values, then Equation (10) can be generalized. The resulting function of ϕ is the *probability distribution function* of the random process $[x(t)]$ at time t_1. This is stated as:

$$P_x(\phi, t_1) = P(x, t_1) = P[x(t_1) \leq \phi] \tag{13}$$

The probability distribution function defines the probability that the instantaneous value of $x(t)$ from a future experiment at time t_1 will be less than or equal to the amplitude ϕ of interest. The general probability distribution function is illustrated in Figure 3-6.

For ergodic (stationary) data, the probability distribution function is the same at all times and can be computed from a single measurement of $x(t)$:

$$P(x) = P[x(t) \leq \phi] = \underset{T \rightarrow \infty}{\text{lim}} \frac{T[x(t) \leq \phi]}{T} \tag{14}$$

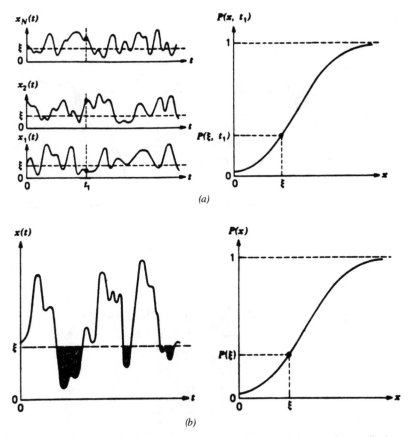

FIGURE 3-6. Probability distribution function: (a) general case; (b) function for ergodic data.

Note that $T[x(t) \leq \phi]$ is the total time that process $x(t)$ is less than or equal to the amplitude ϕ (see Figure 3-6). For this special case, the probability distribution function defines the probability that the instantaneous value of $x(t)$ from a future experiment at any time is $< \phi$.

Note that the probability distribution function $P(x)$ exists between the limits of zero and unity. It is common to describe the behavior of $P(x)$ over these limits in terms of the slope of the functions. The slope is simply the derivative of the distribution function; $p(x,t_1) = [dP(x,t_1)]/dx$.

The resulting function is called the *probability density* function and for stationary data there is no time dependence, i.e.:

$$p(x,t_1) \frac{dP(x,t_1)}{dx} = p(x) \qquad (15)$$

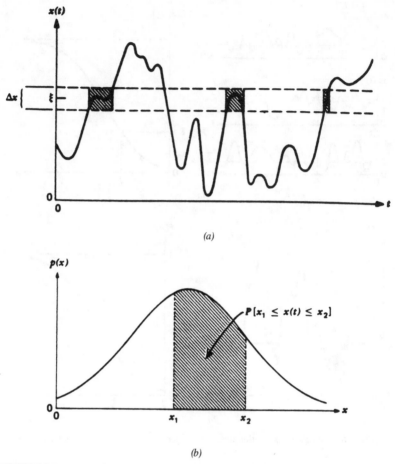

(a)

(b)

FIGURE 3-7. Probability density function: (a) probability density measurement; (b) probability calculation from the density function.

The probability density function is illustrated in Figure 3-7. In the figure, a narrow amplitude band Δx centered at ϕ is chosen. For stationary (ergodic) data, the probability that $x(t)$ will fall within this chosen band at any arbitrary time is:

$$P[x(t) \in \Delta x_\phi] = \lim_{T \to \infty} \frac{T[x(t) \in \Delta x_\phi]}{T} \tag{16}$$

where symbol $\in$ means "within." Hence, $T[x(t) \in \Delta x_\phi]$ refers to the time $x(t)$ falls within the band Δx centered at ϕ. The probability density function is computed as the rate of change of probability versus amplitude:

$$p(x) = \lim_{\Delta x \to 0} \frac{P[x(t) \in \Delta x_\phi]}{\Delta x} \tag{17}$$

The probability of occurrence is generated by integrating the probability density function; i.e., by calculating the area under $p(x)$ between two specified amplitudes (refer to Figure 3-7b). Hence, the probability of finding an amplitude in the range between x_1 and x_2 is:

$$P[x_1 < x(t) \leq x_2] = \int_{x_1}^{x_2} p(x)\, dx = P(x_2) - P(x_1) \tag{18}$$

or

$$P[-\infty < x(t) \leq x_2] = \int_{-\infty}^{x_2} p(x)\, dx = P(x) \tag{19}$$

where $x \to -\infty$.

Equation (19) states that the area under the probability density function below the amplitude x_2 is the value of the probability distribution function evaluated at x_2. And when $x_2 \to \infty$ this reduces to the following:

$$P[-\infty < x(t) \leq \infty] = \int_{-\infty}^{\infty} p(x)\, dx = 1 \tag{20}$$

That is, the total area under the probability density function is unity, meaning simply that the amplitude falls somewhere between $\pm \infty$ at any time.

We can now introduce several definitions which describe the statistical properties of data. These definitions will help to quantify average properties that describe the central tendency and spread or dispersion of data.

The first of these terms are *expected values* and *moments*. For any single-valued function $g(x)$, the expected value of $g(x)$ is:

$$E[g(x)] = \int_{-\infty}^{\infty} g(x)\, p(x)\, dx \tag{21}$$

Since Equation (21) is linear, the additivity rule applies:

$$E[g(x) + h(x)] = E[g(x)] + E[h(x)] \tag{22}$$

And it is also a homogeneous function, hence:

$$E[cg(x)] = cE[g(x)] \tag{23}$$

Moments for a stationary random process are defined as:

$$\mu_k = E[x^k] = \int_{-\infty}^{\infty} x^k p(x) dx, \text{ where } k = 0,1,2, \ldots \tag{24}$$

where $p(x)$ = probability density function of a stationary random process $\{x(t)\}$; μ_k = kth moment.

For the zero moment ($k = 0$):

$$\mu_0 = E[x0] = \int_{-\infty}^{\infty} x^0 p(x) \, dx = 1 \tag{25}$$

For the first moment ($k = 1$):

$$\mu_1 = E[x1] = \int_{-\infty}^{\infty} x p(x) \, dx = \mu \tag{26}$$

For the second moment ($k = 2$):

$$\mu_2 = E[x2] = \int_{-\infty}^{\infty} x^2 p(x) \, dx = \psi^2 \tag{27}$$

and so on.

The first and second moments are generally the most significant in data analysis. The first moment, μ, is called the *mean value* of $\{x(t)\}$, and the second moment, ψ^2 is called the *mean square value* of $\{x(t)\}$. The positive square root of the mean square value is referred to as the rms value (i.e., root mean square value).

When concerned with the higher order moments ($k > 2$), it is common practice to compute the moments about the mean. These are referred to as *central moments*.

The *variance* is the second second central moment:

$$\mu_2^c = E[(x - \mu)^2] = \int_{-\infty}^{\infty} (x - \mu)^2 \, p(x) \, dx = \sigma^2 \qquad (28)$$

Recall that the *standard deviation* is defined as the positive square root of the variance.

The significance of μ and σ^2 are as follows. The first moment (mean value) quantifies the central tendency of a random process, whereas the variance defines the process' dispersion. The second moment, ψ^2, gives a measure of both the central tendency and dispersion or spread of data; that is, they are interrelated as follows:

$$\psi^2 = \sigma^2 + \mu^2 \qquad (29)$$

Equations (26) through (28) can be expressed more practically in terms of the time history data representative of the random process. For the most general case of non-stationary data, the mean value at time t_1 is:

$$\mu(t_1) = E[x(t_1)] = \lim_{N \to \infty} \frac{1}{N} \sum_{i=1}^{N} x_i(t_1) \qquad (30)$$

$$\psi^2(t_1) = E[x^2(t_1)] = \lim_{N \to \infty} \frac{1}{N} \sum_{i=1}^{N} x_i^2(t_1) \qquad (31)$$

$$\sigma^2(t_1) = E[\{x(t_1) = \mu(t_1)\}^2] = \lim_{N \to \infty} \frac{1}{N} \sum_{i=1}^{N} \{x_i(t_1) - \mu(t_1)\}^2 \qquad (32)$$

For stationary (ergodic) data, the various moments are constants which are independent of time. These can be calculated in a straightforward manner from a single time history measurement of $x(t)$:

$$\mu = E[x(t)] = \lim_{T \to \infty} \frac{1}{T} \int_{o}^{T} x(t) dt \qquad (33)$$

$$\psi^2 = E[x^2(t)] = \lim_{T \to \infty} \frac{1}{T} \int_{o}^{T} x^2(t) dt \qquad (34)$$

$$\sigma^2 = E[\{x(t) - \mu\}^2] = \lim_{T \to \infty} \frac{1}{T} \int_0^T \{x(t) - \mu\}^2 dt \qquad (35)$$

The general shape of the probability density function cannot be predicted a priori for an experiment. There are, however, several probability density functions that are widely observed in nature and which have been rigorously defined through mathematics. Two of these probability density functions are Gaussian and sinusoidal. Many random physical phenomena can be reasonably well approximated by a normal distribution (Gaussian):

$$p(x) = \frac{1}{\sigma\sqrt{2\pi}} \exp[-(x - \mu)^2/2\sigma^2] \qquad (36)$$

where μ = mean value; σ = standard deviation.

For purposes of tabulating values of the Gaussian distribution it is convenient to use a standardization variable:

$$z = \frac{(x - \mu)}{\sigma}$$

where z has zero mean and unity standard deviation. Substituting this transformation variable into Equation (36) gives:

$$p(z) = \frac{1}{\sqrt{2\pi}} e^{-z^2/2} \qquad (37)$$

Equation (37) is the standardized Gaussian probability density function. The standardized normal probability distribution function is:

$$P(z) = \frac{1}{\sqrt{2\pi}} \int_{-\infty}^{z} P(\phi)d\phi \qquad (38)$$

It should be noted that most physical phenomena are comprised of a net number of constituent random events; hence, a normal distribution often provides a fair approximation to the probability density function of random data. However, the Gaussian assumption is only valid for random data of interest when no deterministic components are present.

Deterministic data have repetitive or periodic amplitudes that can be decomposed into a collection of harmonically related sine waves. For a single sine wave, the exact amplitude at any future time is $x(t) = X\sin(2\pi ft + \theta)$.

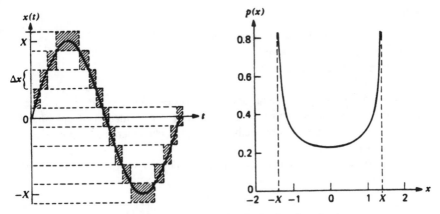

FIGURE 3-8. Standardized probability density function of a sine wave.

We can, however, view the phase angle θ as a random variable having uniform distribution over the limits of $\pm \pi$. A sinusoidal function can then be considered a random process where its probability density function is:

$$P(x) = (\pi\sqrt{2\sigma^2 - x^2})^{-1} \quad , \quad |x| < X$$
$$= 0 \qquad\qquad\qquad , \quad |x| \geq X \tag{39}$$

where the standard deviation of the sine wave is $\sigma = X/\sqrt{2}$.

Figure 3-8 shows a plot of the probability density function standardized for $\sigma = 1$. The probability density function of a sine wave is also determined from the mean and standard deviation of the data. In contrast to the Gaussian form, however, the probability density of a sinusoidal signal reaches a minimum at the mean value of the sine wave. This means that a value at the mean is the least likely event between the limits of $\pm X$. For a Gaussian form, a value at the mean is the most likely event. Hence, there is a significant difference between a sine wave and a narrow-band noise that is approximately Gaussian.

A final case to consider is that of a sine wave in Gaussian noise. Consider a stationary random time history of the form $x(t) = n(t) + s(t)$, where $n(t)$ represents Gaussian random noise, and the sine wave is $s(t) = S \sin (2\pi ft + \theta)$. The probability density function of process $x(t)$ is the convolution integral of the individual density functions. For zero mean values for $n(t)$ and $s(t)$, the probability density is:

$$p(x) = \frac{1}{\sigma_n \pi \sqrt{2\pi}} \int_0^\pi \exp\left[-\left(\frac{x - S\cos\theta}{4\sigma_n}\right)^2\right] d\theta \tag{40}$$

where

σ_n = standard deviation of Gaussian noise $n(t)$
S = amplitude of sine wave
θ = phase angle of sine wave

As noted earlier, spectral analysis of time series data takes place in the frequency domain, and is performed using Fourier transformations. Generalized Fourier series representations enable data to be characterized as points in a vector space.

Any periodic function can be expanded in a Fourier series using the following formula:

$$x(t) = \frac{a_o}{2} + \sum_{k=1}^{\infty} (a_k \cos 2\pi f_k t + b_k \sin 2\pi f_k t) \tag{41}$$

where the fundamental frequency is:

$$f_k = kf_1 = \frac{k}{T}, \qquad k = 1, 2, 3, \ldots \tag{42}$$

Equation (41) states that any periodic function $x(t)$ can be expressed in terms of sine and cosine waves at discrete frequencies. Coefficients $\{a_k\}$ and $\{b_k\}$ are calculated from the following integrals:

$$a_k = \frac{2}{T} \int_0^T x(t) \cos 2\pi f_k t \, dt, \qquad k = 0, 1, 2, \ldots \tag{43}$$

$$b_k = \frac{2}{T} \int_0^T x(t) \sin 2\pi f_k t \, dt, \qquad k = 1, 2, 3, \ldots \tag{44}$$

The mean value of $x(t)$ is:

$$\mu_x = \frac{a_o}{2} = \frac{1}{T} \int_0^T x(t) \, dt \tag{45}$$

The above equations can be found in any standard statistics textbook but are sometimes expressed in equivalent forms using $\omega = 2\pi f$.

Other forms of the Fourier series expansion are as follows:

$$x(t) = X_o + \sum_{k=1}^{\infty} X_k \cos(2\pi f_k t - \theta_k) \tag{46}$$

$$x(t) = \sum_{k=-\infty}^{\infty} A_k e^{j2\pi f_k t} \tag{47}$$

Equation (46) expresses $x(t)$ in polar form, and the various parameters are defined as follows: $\{X_k\}$ = amplitude factors = $\sqrt{a_k^2 + b_k^2}$ with $k = 1,2,3,\ldots$; $\{\theta_k\}$ = phase factors = $\tan^{-1}(b_k/a_k)$; $X_o = a_o/2$; f_k = discrete frequencies. In Equation (47) the following definitions are included:

$$A_o = a_o/2;$$

$$A_k = \frac{1}{2}(a_k - jb_k) = \frac{1}{T}\int_0^T x(t)\, e^{-j2\pi f_k t} dt$$

where $k = \pm 1, \pm 2, \pm 3, \ldots$

Note that $x(t)$ can be expressed in a complex-valued form using both negative and positive frequency components. Although physically this may not occur, at least mathematically it is valid. Hence, the factors will be complex valued with:

$$A_k = |A_k| e^{-j\theta}k, \; k = \pm 1, \pm 2, \pm 3, \ldots \tag{48}$$

$$|A_k| = \frac{1}{2}\sqrt{a_k^2 + b_k^2} = \frac{X_k}{2} \tag{49}$$

$$\theta_k = \tan^{-1}(b_k/a_k) \tag{50}$$

The above relationships are readily derived from Euler's theorem ($e^{-j\theta} = \cos\theta - j\sin\theta$).

For a real-valued $x(t)$ function, the factors are expressed in terms of their complex conjugates. For example, $|A_{-k}| = |A_k|$ for $\theta_{-k} = -\theta_k$ and $A_{-k} = |A_{-k}| e^{-j\theta}k = |A_k| e^{j\theta}k = A_k^*$; where A_k^* is the complex conjugate of A_k.

When the period T of function $x(t)$ approaches infinity, a Fourier integral is obtained. This is expressed as:

$$X(f) = \int_{-\infty}^{\infty} x(t)e^{-j2\pi ft}dt \qquad (51)$$

valid for $-\infty < f < \infty$.

$$X(f) \text{ exists when } \int_{-\infty}^{\infty} |x(t)|\,dt < \infty.$$

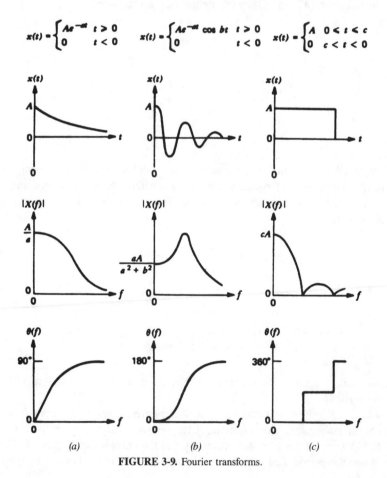

FIGURE 3-9. Fourier transforms.

Table 3-4. Examples of Fourier Transform Pairs.

$x(t)$	$X(f)$
1	$\delta(f)$
$e^{j2\pi f_0 t}$	$\delta(f - f_0)$
$x(t - \tau_0)$	$X(f)e^{-j2\pi f \tau_0}$
$\cos 2\pi f_0 t$	$\frac{1}{2}[\delta(f - f_0) + \delta(f + f_0)]$
$\sin 2\pi f_0 t$	$\frac{1}{2j}[\delta(f - f_0) - \delta(f + f_0)]$
$1; \ 0 \le t \le T$ $0; \ \text{otherwise}$	$T\left(\frac{\sin \pi f t}{\pi f t}\right)e^{-j\pi f T}$
$2aB\left(\frac{\sin 2\pi B t}{2\pi B t}\right)$	$a; \ -B \le f \le B$ $0; \ \text{otherwise}$
$aB\left(\frac{\sin \pi B t}{\pi B t}\right)\cos 2\pi f_0 t$	$a; f_0 - \dfrac{B}{2} \le f \le f_0 + \dfrac{B}{2}$. $0; \ \text{otherwise}$
$e^{-a|t|}; \ a > 0$	$\dfrac{2a}{a^2 + (2\pi f)^2}$
$e^{-a|t|}\cos 2\pi f_0 t; \ a > 0$	$\dfrac{a}{a^2 + 4\pi^2(f + f_0)^2} + \dfrac{a}{a^2 + 4\pi^2(f - f_0)^2}$
$\displaystyle\int_{-\infty}^{\infty} x_1(u)\, x_2(t - u)\, du$	$X_1(f)\, X_2(f)$

$X(f)$ is referred to as the direct Fourier transform of $x(t)$, or in signal analysis, the *spectrum*. The inverse Fourier transform of $X(f)$ gives $x(t)$:

$$x(t) = \int_{-\infty}^{\infty} X(f)e^{j2\pi f t}dt, \ -\infty < t < \infty \quad (52)$$

Examples of Fourier spectra for different time history records are illustrated in Figure 3-9. Special Fourier transform pairs are given in Table 3-4.

Spectral analysis as described here is best applied to developing the power spectrum (i.e. a plot of the response variance as a function of frequency). This enables dominant frequencies in the system to be identified, from which specific independent variables causing these frequencies can be found. In a later chapter we will return to this subject to establish a basis for cross-correlating events in a process.

SAMPLING ERROR

In all the formulas presented on probability functions the limits of the integrals were $\pm \infty$. In practice, however, it is impossible to analyze an infinite number of data sets or even a single record of infinite length. Consequently, all experiments have statistical sampling errors that result from their analyses. The result of any analysis of random data produces only a sample estimate of the specific parameter under investigation. As noted earlier, the magnitude of the sample estimate error has a direct impact on interpretations and the degree of confidence in applying the analysis.

There are basically two types of errors associated with random data analysis. The first of these is random error which results from averaging operations performed over a finite data set. Whenever averaging of a finite number of sample records or over a finite time length is performed, a random error is generated.

The second type is more of a systematic error that consistently appears from one analysis to another. This is referred to as biased error. It is most often associated with calculations or derivatives—i.e., where increments of the amplitude are used to transform a probability estimate to a probability density estimate. Note that the theoretical expression of $p(x)$ in Equation (17) is based on the limit as $\Delta x \rightarrow 0$; however, in an actual analysis Δx must be finite.

For a parameter ξ to be estimated from independent sample records to produce a collection of estimate $\hat{\xi}_i$ (where $i = 1,2,3,...$), the bias error of the estimate $\hat{\xi}$ is:

$$\text{Bias Error} = b_{\hat{\xi}} = E[\hat{\xi}] - \xi = \lim_{N \to \infty} \frac{1}{N} \sum_{i=1}^{N} \hat{\xi}_i - \xi \tag{53}$$

The random error of the estimate $\hat{\xi}$ is:

$$\text{Random Error} = \sigma_{\xi} = \left[\lim_{N \to \infty} \frac{1}{N} \sum_{i=1}^{N} \{\hat{\xi}_i - E[\hat{\xi}]\}^2 \right]^{1/2} \tag{54}$$

Equation (54) is the standard deviation of the estimate about its expected value.

When handling large quantities of data it is convenient to express errors in terms of normalized quantities:

$$\epsilon_b = \frac{b_{\hat{\xi}}}{\xi} ; \qquad \epsilon_r = \frac{\sigma_{\hat{\xi}}}{\xi} \tag{55}$$

The normalized random error ϵ_r is given the name the coefficient of variation of the estimate. For example, if $\epsilon_r = 0.15$, the scatter in $\hat{\xi}$ about the average estimate would have a standard deviation of 10% of the ξ value. Also, for $\epsilon_b = 0.15$, the estimate $\hat{\xi}$ is typically 15% greater than ξ.

The sampling distribution of the estimate is the probability density function. As a general guideline, when $\epsilon_r < 0.2$, the sampling distribution can be reasonably well approximated by a Gaussian distribution. In this case the mean value is:

$$\mu_{\hat{\xi}} = (1 + \epsilon_b)\xi$$

and the standard deviation is:

$$\sigma_{\hat{\xi}} = \epsilon_r \xi$$

Consequently, the probability can be approximated by:

$$p(\hat{\xi}) \approx \frac{1}{\epsilon_r \xi \sqrt{2\pi}} \exp\left\{ \frac{-[\hat{\xi} - (1 + \epsilon_b)\xi]^2}{2(\epsilon_r \xi)^2} \right\} \tag{56}$$

Probability density estimates will contain both bias and random error. As a first order approximation, the bias error in the estimate $\hat{p}(x)$ can be computed from:

$$\epsilon_b[\hat{p}(x)] \cong \frac{(\Delta x)^2}{24} \frac{p''(x)}{p(x)} \tag{57}$$

where $p''(x)$ denotes the second derivative (i.e., $= [d^2p(x)]/dx^2$).

When $p(x)$ is estimated by ensemble averaging over statistically independent time history records, the associated random error can be estimated from:

$$\epsilon_r[\hat{p}(x)] \approx \frac{1}{\sqrt{N} \, \Delta x \, p(x)} \tag{58}$$

where N is the number of records.

When time averaging over a single record, the random error associated with the $\hat{p}(x)$ estimated can be approximated by:

$$\epsilon_r[\hat{p}(x)] \approx \frac{1}{\sqrt{2BT\Delta x} \, p(x)} \tag{59}$$

where T = length of record; B = data bandwidth (Hz). Equation (59) assumes the idealized case that the energy of the data is uniformly distributed over the bandwidth.

There are no bias errors in the estimate of mean and mean square values. The random errors in these estimates are well defined and are summarized below:

For Ensemble Averaging:

$$\hat{\mu}_x = \frac{1}{N}\sum_{i=1}^{N} x_i; \quad \epsilon_r = \frac{\sigma_x}{\mu_x\sqrt{N}}\text{(mean value estimate)} \tag{60}$$

$$\hat{\psi}_x^2 = \frac{1}{N}\sum_{i=1}^{N} x_i^2; \quad \epsilon_r = \sqrt{2/N}\text{(mean square value estimate)} \tag{61}$$

For Time Averaging:

$$\hat{\mu}_x \cong \frac{1}{T}\int_0^T x(t)dt; \quad \epsilon_r \cong \frac{\sigma_x}{\mu_x\sqrt{2BT}} \text{ (mean value estimate)} \tag{62}$$

$$\hat{\psi}_x^2 \cong \frac{1}{T}\int_0^T x^2(t)dt; \quad \epsilon_r \cong \frac{1}{\sqrt{BT}} \text{ (mean square value estimate)} \tag{63}$$

Formulas (60) and (61) for ensemble averaging may be considered as exact. For the time average estimates however, Equations (62) and (63), the error formulas are based on the idealized assumption that the energy of the data is uniformly distributed in frequency over a bandwidth of B. From statistical sampling theory this means that $N \cong 2BT$, which is generally not the case. Hence, Equations (62) and (63) should only be used for order of magnitude estimates.

Only errors that arise from statistical sampling procedures have been considered. There are many other potential sources of error in an experimental program. These errors generally arise during the data acquisition stages and depend on the specific measurement scheme and the nature of the experiment.

PRACTICE PROBLEMS

1 Table 3-5 reports moisture levels during the production of rubber bales.
   (A) Determine whether the process is in a state of statistical control.
   (B) Construct a Cusum chart and indicate how many shifts occurred in the process average and at what runs they occurred.

Table 3-5. Moisture Levels (wt. %).

Lot	Water (%)	Lot	Water (%)
1	1.5	16	1.0
2	2.0	17	1.8
3	2.2	18	6.1
4	2.5	19	3.1
5	3.0	20	3.2
6	4.0	21	3.7
7	7.0	22	4.5
8	2.0	23	4.6
9	2.5	24	3.0
10	7.3	25	2.7
11	5.2	26	2.6
12	3.5	27	2.9
13	0.5	28	2.1
14	0.5	29	2.2
15	1.0	30	2.7

2 Pressure fluctuations in a fluidized bed reactor are measured with a piezoelectric transducer. Values of the actual signal reading in millivolts, obtained at 6 second intervals are listed below. Can you deduce from the statistical analysis whether bubbles are passing by the transducer and if so, at what mean frequency?

50 mv	69	60	67
55	50	53	60
75	57	47	48
40	60	57	47
43	79	83	55
69	61	72	75
57	45	61	73
53	43	72	60
37	42	67	55
52	87	50	50
59	83	51	50
76	70	50	52
77	55	71	79
53	55	61	67
50	54	62	59
47	69	50	58

3 Prepare a plot of the standardized probability density versus the standardized amplitude using Equation (40).

CHAPTER 4

Analysis of Variance

INTRODUCTION

The method of analysis of variance is used to separate the effects of qualitative factors, or classifications, on an observed variable of interest. As an example, we may want to evaluate the quality of two raw feedstock materials as measured by two different analytical techniques. This statistical method enables us to determine the variance in quality attributed to the difference between methods as well as the difference between the feedstocks. As explained later in this chapter, it is a powerful technique for structuring factorial experiments. That is, we may statistically design an experimental program to study the key phenomenon of interest within a defined limit of confidence in our conclusions. This chapter illustrates the mechanics of several simple statistical tests for evaluating the statistical significance of data and the conclusions derived from them. The method of analysis of variance is outlined and demonstrated on several practical examples.

PROCEDURES FOR COMPARING MEANS AND VARIANCES

By way of review, a variance (s^2) is an averaged-out sum-of-squared deviations. For sample size n, the total variance is:

$$\sigma^2 = \frac{\Sigma(x_i - \bar{x}_i)^2}{(n - 1)} \tag{1}$$

where $(n - 1)$ is the number of degrees of freedom and $\bar{x}$ the sample mean. Equation (1) can also be written as:

$$\sigma^2 = \frac{\Sigma(x_i^2) - n\bar{x}^2}{(n - 1)} \tag{2}$$

An important point to realize is that the total variance on the sum-of-the-squared-deviations may in part be comprised of random errors as well as other factors. The analysis of variance is applied to determining the importance of any factors which seemingly affect the process. The factor (or

71

factors) is considered significant if the variance it explains is substantially greater than the variance it leaves unexplained. The test of significance used is called the F-test which relates to the ratio of variances. The analysis is capable of handling three or more factors at a time.

Before reviewing the procedures, we briefly describe how analysis of variance works. Probability statements in the analysis of variance utilize the ratios of the mean squared deviations to the residual mean square deviations. Since the mean squared deviation values are variances, the ratios are F-ratios.

F-values are used to determine the probability of two variances being from the same population. When the probability of this occurring is small, we conclude that they are not from the same population. Each variance relates to a control parameter and thus provides an indication of the relative size of the parameter's effect on the dependent variable under consideration.

The relative effect in itself does not provide sufficient information. An important question to answer is what is the importance of a control parameter as compared with some objective standard? In addition, it is important to know whether there are important variations that have not been considered. Normally, the objective standard to use is the natural variation of the dependent variable under steady-state conditions.

The natural variation is established by small, random changes in control variables which are referred to as *noise*. When signals are overlaid with noise, they are indistinguishable from it. Sometimes there are several random factors which contribute to the make-up of noise, but note that the sum of their contributions will approach zero. Despite this fact, they will contribute some variance. The F-ratio between the signal variance and the noise variance in this situation provides an indication of whether the parameter is significant or not. Specifically, a large F-ratio ($F > 1$) indicates that the parameter is significant, whereas small F indicates that it is contributing only to the system noise.

Statistical significance is established by the level of probability associated with the F-value. When the background noise level is low but steady, control parameters that in fact are weak may show up with a high degree of statistical significance. It is important, therefore, to be able to distinguish between statistical and practical significance. By statistical significance, it is meant that the effect of a parameter on a process can be distinguished from the system noise. Fortunately, in some measurement systems, signals or data unrelated to the parameter of interest can be readily identified and discarded during measurement. Examples of measurement systems used in fluid dynamic studies where this is practical are given by Cheremisinoff (1986). Unfortunately, this is only one area of application, and more often steady-state noise cannot be readily separated from the unsteady *residual variance*. In the design of experiments and measurement systems, options for separating out identifiable noise should always be considered. If this is not feasible, then the noise will be masked in the residual variance and that term is employed in the F-test.

Detailed discussions can be found in the volumes by Atchley and Bryant (1975).

There are basically two sources of the residual variance. The first is comprised of short- and long-term variation attributed to environmental aspects, e.g., ambient temperature fluctuations, raw material changes, electrical surges in instrumentation, instrument drift. The second source relates to the inherent inaccuracy of the measurement at a fixed point in time. Single measurements obtained over long time intervals such as a period of days, weeks or months result in a population sample in which the steady-state and residual variances cannot be separated. In this situation the combined variance must be used in the F-test.

Theoretically, we could measure the true system noise by repeated measurements with no lag time in between. In practice, we can only estimate the true system noise by repeating the measurements as close together as possible. The value of having an independent approximation of the system noise is that the F-test can then be applied to assess whether the residual variance is indeed significant. If this is the case, then we have failed to identify at least one factor affecting the system. This is illustrated later in this chapter.

Two-Sided Comparison of Two Variances

This procedure enables one to test for the equality of two population variances, σ_1^2 and σ_2^2. A two-sided hypothesis is used:

$$H_o: \sigma_1^2 = \sigma_2^2, \text{ or}$$

$$H_o: \sigma_1^2/\sigma_2^2 = 1$$

In this analysis we are concerned with both the alternatives of $\sigma_1^2 > \sigma_2^2$ and $\sigma_1^2 < \sigma_2^2$.

The specific analysis procedure is as follows:

- Select an α-risk (i.e., the risk of asserting a difference when none exists).
- Compute the variance of the two samples, s_1^2 and s_2^2 with $\nu_1 = n_1 - 1$ and $\nu_2 = n_2 - 1$, degrees of freedom.
- Table 4-1 provides the F-distribution for a two-sided comparison of variance. Obtain a value for F for $100(1-\alpha)\%$ confidence and ν_N, ν_D degrees of freedom. Note that for:

$$s_1^2 > s_2^2, \nu_N = n_1 - 1, \nu_D = n_2 - 1$$

$$s_2^2 > s_1^2, \nu_N = n_2 - 1, \nu_D = n_1 - 1$$

Table 4-1. Two sided F-Distribution.

α = 0.10

(continued)

ν_D \ ν_N	1	2	3	4	5	6	7	8	9	10	12	15	20	24	30	40	60	120	∞
1	161.4	199.5	215.7	224.6	230.2	234.0	236.8	238.9	240.5	241.9	243.9	245.0	248.0	249.1	250.1	251.1	252.2	253.3	254.3
2	18.51	19.00	19.16	19.25	19.30	19.33	19.35	19.37	19.38	19.40	19.41	19.43	19.45	19.45	19.46	19.47	19.48	19.49	19.50
3	10.13	9.55	9.28	9.12	9.01	8.94	8.89	8.85	8.81	8.79	8.74	8.70	8.66	8.64	8.62	8.59	8.57	8.55	8.53
4	7.71	6.94	6.59	6.39	6.26	6.16	6.09	6.04	6.00	5.96	5.91	5.86	5.80	5.77	5.75	5.72	5.69	5.66	5.63
5	6.61	5.79	5.41	5.19	5.05	4.95	4.88	4.82	4.77	4.74	4.68	4.62	4.56	4.53	4.50	4.46	4.43	4.40	4.36
6	5.99	5.14	4.76	4.53	4.39	4.28	4.21	4.15	4.10	4.06	4.00	3.94	3.87	3.84	3.81	3.77	3.74	3.70	3.67
7	5.59	4.74	4.35	4.12	3.97	3.87	3.79	3.73	3.68	3.64	3.57	3.51	3.44	3.41	3.38	3.34	3.30	3.27	3.23
8	5.32	4.46	4.07	3.84	3.69	3.58	3.50	3.44	3.39	3.35	3.28	3.22	3.15	3.12	3.08	3.04	3.01	2.97	2.93
9	5.12	4.26	3.86	3.63	3.48	3.37	3.29	3.23	3.18	3.14	3.07	3.01	2.94	2.90	2.86	2.83	2.79	2.75	2.71
10	4.96	4.10	3.71	3.48	3.33	3.22	3.14	3.07	3.02	2.98	2.91	2.85	2.77	2.74	2.70	2.66	2.62	2.58	2.54
11	4.84	3.98	3.59	3.36	3.20	3.09	3.01	2.95	2.90	2.85	2.79	2.72	2.65	2.61	2.57	2.53	2.49	2.45	2.40
12	4.75	3.89	3.49	3.26	3.11	3.00	2.91	2.85	2.80	2.75	2.69	2.62	2.54	2.51	2.47	2.43	2.38	2.34	2.30
13	4.67	3.81	3.41	3.18	3.03	2.92	2.83	2.77	2.71	2.67	2.60	2.53	2.46	2.42	2.38	2.34	2.30	2.25	2.21
14	4.60	3.74	3.34	3.11	2.96	2.85	2.76	2.70	2.65	2.60	2.53	2.46	2.39	2.35	2.31	2.27	2.22	2.18	2.13
15	4.54	3.68	3.29	3.06	2.90	2.79	2.71	2.64	2.59	2.54	2.48	2.40	2.33	2.29	2.25	2.20	2.16	2.11	2.07
16	4.49	3.63	3.24	3.01	2.85	2.74	2.66	2.59	2.54	2.49	2.42	2.35	2.28	2.24	2.19	2.15	2.11	2.06	2.01
17	4.45	3.59	3.20	2.96	2.81	2.70	2.61	2.55	2.49	2.45	2.38	2.31	2.23	2.19	2.15	2.10	2.06	2.01	1.96
18	4.41	3.55	3.16	2.93	2.77	2.66	2.58	2.51	2.46	2.41	2.34	2.27	2.19	2.15	2.11	2.06	2.02	1.97	1.92
19	4.38	3.52	3.13	2.90	2.74	2.63	2.54	2.48	2.42	2.38	2.31	2.23	2.16	2.11	2.07	2.03	1.98	1.93	1.88
20	4.35	3.49	3.10	2.87	2.71	2.60	2.51	2.45	2.39	2.35	2.28	2.20	2.12	2.08	2.04	1.99	1.95	1.90	1.84
21	4.32	3.47	3.07	2.84	2.68	2.57	2.49	2.42	2.37	2.32	2.25	2.18	2.10	2.05	2.01	1.96	1.92	1.87	1.81
22	4.30	3.44	3.05	2.82	2.66	2.55	2.46	2.40	2.34	2.30	2.23	2.15	2.07	2.03	1.98	1.94	1.89	1.84	1.78
23	4.28	3.42	3.03	2.80	2.64	2.53	2.44	2.37	2.32	2.27	2.20	2.13	2.05	2.01	1.96	1.91	1.86	1.81	1.76
24	4.26	3.40	3.01	2.78	2.62	2.51	2.42	2.36	2.30	2.25	2.18	2.11	2.03	1.98	1.94	1.89	1.84	1.79	1.73
25	4.24	3.39	2.99	2.76	2.60	2.49	2.40	2.34	2.28	2.24	2.16	2.09	2.01	1.96	1.92	1.87	1.82	1.77	1.71
26	4.23	3.37	2.98	2.74	2.59	2.47	2.39	2.32	2.27	2.22	2.15	2.07	1.99	1.95	1.90	1.85	1.80	1.75	1.69
27	4.21	3.35	2.96	2.73	2.57	2.46	2.37	2.31	2.25	2.20	2.13	2.06	1.97	1.93	1.88	1.84	1.79	1.73	1.67
28	4.20	3.34	2.95	2.71	2.56	2.45	2.36	2.29	2.24	2.19	2.12	2.04	1.96	1.91	1.87	1.82	1.77	1.71	1.65
29	4.18	3.33	2.93	2.70	2.55	2.43	2.35	2.28	2.22	2.18	2.10	2.03	1.94	1.90	1.85	1.81	1.75	1.70	1.64
30	4.17	3.32	2.92	2.69	2.53	2.42	2.33	2.27	2.21	2.16	2.09	2.01	1.93	1.89	1.84	1.79	1.74	1.68	1.62
40	4.08	3.23	2.84	2.61	2.45	2.34	2.25	2.18	2.12	2.08	2.00	1.92	1.84	1.79	1.74	1.69	1.64	1.58	1.51
60	4.00	3.15	2.76	2.53	2.37	2.25	2.17	2.10	2.04	1.99	1.92	1.84	1.75	1.70	1.65	1.59	1.53	1.47	1.39
120	3.92	3.07	2.68	2.45	2.29	2.17	2.09	2.02	1.96	1.91	1.83	1.75	1.66	1.61	1.55	1.50	1.43	1.35	1.25
∞	3.84	3.00	2.60	2.37	2.21	2.10	2.01	1.94	1.88	1.83	1.75	1.67	1.57	1.52	1.46	1.39	1.32	1.22	1.00

Table 4.1. (continued).

α = 0.05

$\nu_D \backslash \nu_N$	1	2	3	4	5	6	7	8	9	10	12	15	20	24	30	40	60	120	∞
1	647.8	799.5	864.2	899.6	921.8	937.1	948.2	956.7	963.3	968.6	976.7	984.9	993.1	997.2	1001	1006	1010	1014	1018
2	38.51	39.00	39.17	39.25	39.30	39.33	39.36	39.37	39.39	39.40	39.41	39.43	39.45	39.46	39.46	39.47	39.48	39.49	39.50
3	17.44	16.04	15.44	15.10	14.88	14.73	14.62	14.54	14.47	14.42	14.34	14.25	14.17	14.12	14.08	14.04	13.99	13.95	13.90
4	12.22	10.65	9.98	9.60	9.36	9.20	9.07	8.98	8.90	8.84	8.75	8.66	8.56	8.51	8.46	8.41	8.36	8.31	8.26
5	10.01	8.43	7.76	7.39	7.15	6.98	6.85	6.76	6.68	6.62	6.52	6.43	6.33	6.28	6.23	6.18	6.12	6.07	6.02
6	8.81	7.26	6.60	6.23	5.99	5.82	5.70	5.60	5.52	5.46	5.37	5.27	5.17	5.12	5.07	5.01	4.96	4.90	4.85
7	8.07	6.54	5.89	5.52	5.29	5.12	4.99	4.90	4.82	4.76	4.67	4.57	4.47	4.42	4.36	4.31	4.25	4.20	4.14
8	7.57	6.06	5.42	5.05	4.82	4.65	4.53	4.43	4.36	4.30	4.20	4.10	4.00	3.95	3.89	3.84	3.78	3.73	3.67
9	7.21	5.71	5.08	4.72	4.48	4.32	4.20	4.10	4.03	3.96	3.87	3.77	3.67	3.61	3.56	3.51	3.45	3.39	3.33
10	6.94	5.46	4.83	4.47	4.24	4.07	3.95	3.85	3.78	3.72	3.62	3.52	3.42	3.37	3.31	3.26	3.20	3.14	3.08
11	6.72	5.26	4.63	4.28	4.04	3.88	3.76	3.66	3.59	3.53	3.43	3.33	3.23	3.17	3.12	3.06	3.00	2.91	2.88
12	6.55	5.10	4.47	4.12	3.89	3.73	3.61	3.51	3.44	3.37	3.28	3.18	3.07	3.02	2.96	2.91	2.85	2.79	2.72
13	6.41	4.97	4.35	4.00	3.77	3.60	3.48	3.39	3.31	3.25	3.15	3.05	2.95	2.89	2.84	2.78	2.72	2.66	2.60
14	6.30	4.86	4.24	3.89	3.66	3.50	3.38	3.29	3.21	3.15	3.05	2.95	2.84	2.79	2.73	2.67	2.61	2.55	2.49
15	6.20	4.77	4.15	3.80	3.58	3.41	3.29	3.20	3.12	3.06	2.96	2.86	2.76	2.70	2.64	2.59	2.52	2.46	2.40
16	6.12	4.69	4.08	3.73	3.50	3.34	3.22	3.12	3.05	2.99	2.89	2.79	2.68	2.63	2.57	2.51	2.45	2.38	2.32
17	6.04	4.62	4.01	3.66	3.44	3.28	3.16	3.06	2.98	2.92	2.82	2.72	2.62	2.56	2.50	2.44	2.38	2.32	2.25
18	5.98	4.56	3.95	3.61	3.38	3.22	3.10	3.01	2.93	2.87	2.77	2.67	2.56	2.50	2.44	2.38	2.32	2.26	2.19
19	5.92	4.51	3.90	3.56	3.33	3.17	3.05	2.96	2.88	2.82	2.72	2.62	2.51	2.45	2.39	2.33	2.27	2.20	2.13
20	5.87	4.46	3.86	3.51	3.29	3.13	3.01	2.91	2.84	2.77	2.68	2.57	2.46	2.41	2.35	2.29	2.22	2.16	2.09
21	5.83	4.42	3.82	3.48	3.25	3.09	2.97	2.87	2.80	2.73	2.64	2.53	2.42	2.37	2.31	2.25	2.18	2.11	2.04
22	5.79	4.38	3.78	3.44	3.22	3.05	2.93	2.84	2.76	2.70	2.60	2.50	2.39	2.33	2.27	2.21	2.14	2.08	2.00
23	5.75	4.35	3.75	3.41	3.18	3.02	2.90	2.81	2.73	2.67	2.57	2.47	2.36	2.30	2.24	2.18	2.11	2.01	1.97
24	5.72	4.32	3.72	3.38	3.15	2.99	2.87	2.78	2.70	2.64	2.54	2.44	2.33	2.27	2.21	2.15	2.08	2.01	1.94
25	5.69	4.29	3.69	3.35	3.13	2.97	2.85	2.75	2.68	2.61	2.51	2.41	2.30	2.24	2.18	2.12	2.05	1.98	1.91
26	5.66	4.27	3.67	3.33	3.10	2.94	2.82	2.73	2.65	2.59	2.49	2.39	2.28	2.22	2.16	2.09	2.03	1.95	1.88
27	5.63	4.24	3.65	3.31	3.08	2.92	2.80	2.71	2.63	2.57	2.47	2.36	2.25	2.19	2.13	2.07	2.00	1.93	1.85
28	5.61	4.22	3.63	3.29	3.06	2.90	2.78	2.69	2.61	2.55	2.45	2.34	2.23	2.17	2.11	2.05	1.98	1.91	1.83
29	5.59	4.20	3.61	3.27	3.04	2.88	2.76	2.67	2.59	2.53	2.43	2.32	2.21	2.15	2.09	2.03	1.96	1.89	1.81
30	5.57	4.18	3.59	3.25	3.03	2.87	2.75	2.65	2.57	2.51	2.41	2.31	2.20	2.14	2.07	2.01	1.94	1.87	1.79
40	5.42	4.05	3.46	3.13	2.90	2.74	2.62	2.53	2.45	2.39	2.29	2.18	2.07	2.01	1.91	1.88	1.80	1.72	1.61
60	5.29	3.93	3.34	3.01	2.79	2.63	2.51	2.41	2.33	2.27	2.17	2.06	1.94	1.88	1.82	1.74	1.67	1.58	1.48
120	5.15	3.80	3.23	2.89	2.67	2.52	2.39	2.30	2.22	2.16	2.05	1.94	1.82	1.76	1.69	1.61	1.53	1.43	1.31
∞	5.02	3.69	3.12	2.79	2.57	2.41	2.29	2.19	2.11	2.05	1.94	1.83	1.71	1.64	1.57	1.48	1.39	1.27	1.00

(continued)

Table 4-1. (continued).

α = 0.01

ν_D \ ν_N	1	2	3	4	5	6	7	8	9	10	12	15	20	24	30	40	60	120	∞
1	16211	20000	21615	22500	23056	23437	23715	23925	24091	24224	24426	24630	24836	24940	25044	25148	25253	25359	25465
2	198.5	199.0	199.2	199.2	199.3	199.3	199.4	199.4	199.4	199.4	199.4	199.4	199.4	199.5	199.5	199.5	199.5	199.5	199.5
3	55.55	49.80	47.47	46.19	45.30	44.84	44.43	44.13	43.88	43.69	43.39	43.08	42.78	42.62	42.47	42.31	42.15	41.99	41.83
4	31.33	26.28	24.28	23.15	22.48	21.97	21.62	21.33	21.14	20.97	20.70	20.44	20.17	20.03	19.89	19.75	19.61	19.47	19.32
5	22.78	18.31	16.53	15.56	14.94	14.51	14.20	13.96	13.77	13.62	13.38	13.15	12.90	12.78	12.66	12.53	12.40	12.27	12.14
6	18.63	14.54	12.92	12.03	11.46	11.07	10.79	10.57	10.39	10.25	10.03	9.81	9.59	9.47	9.36	9.24	9.12	9.00	8.88
7	16.24	12.40	10.88	10.05	9.52	9.16	8.89	8.68	8.51	8.38	8.18	7.97	7.75	7.65	7.53	7.42	7.31	7.19	7.08
8	14.69	11.04	9.60	8.81	8.30	7.95	7.69	7.50	7.34	7.21	7.01	6.81	6.61	6.50	6.40	6.29	6.18	6.06	5.95
9	13.61	10.11	8.72	7.96	7.47	7.13	6.88	6.69	6.54	6.42	6.23	6.03	5.83	5.73	5.62	5.52	5.41	5.30	5.19
10	12.83	9.43	8.08	7.34	6.87	6.54	6.30	6.12	5.97	5.85	5.66	5.47	5.27	5.17	5.07	4.97	4.86	4.75	4.64
11	12.23	8.91	7.60	6.88	6.42	6.10	5.86	5.68	5.54	5.42	5.24	5.05	4.86	4.76	4.65	4.55	4.44	4.31	4.23
12	11.75	8.51	7.23	6.52	6.07	5.76	5.52	5.35	5.20	5.09	4.91	4.72	4.53	4.43	4.33	4.23	4.12	4.01	3.90
13	11.37	8.19	6.93	6.23	5.79	5.48	5.25	5.08	4.91	4.82	4.61	4.46	4.27	4.17	4.07	3.97	3.87	3.76	3.65
14	11.06	7.92	6.68	6.00	5.56	5.26	5.03	4.86	4.72	4.60	4.43	4.25	4.06	3.96	3.86	3.76	3.66	3.55	3.44
15	10.80	7.70	6.48	5.80	5.37	5.07	4.85	4.67	4.54	4.42	4.25	4.07	3.88	3.79	3.69	3.58	3.48	3.37	3.26
16	10.58	7.51	6.30	5.61	5.21	4.91	4.69	4.52	4.38	4.27	4.10	3.92	3.73	3.61	3.51	3.44	3.33	3.22	3.11
17	10.38	7.35	6.16	5.50	5.07	4.78	4.56	4.39	4.25	4.14	3.97	3.79	3.61	3.51	3.41	3.31	3.21	3.10	2.98
18	10.22	7.21	6.03	5.37	4.96	4.60	4.44	4.28	4.14	4.03	3.86	3.68	3.50	3.40	3.30	3.20	3.10	2.99	2.87
19	10.07	7.09	5.92	5.27	4.85	4.56	4.34	4.18	4.04	3.93	3.76	3.59	3.40	3.31	3.21	3.11	3.00	2.89	2.78
20	9.91	6.99	5.82	5.17	4.76	4.47	4.26	4.09	3.96	3.85	3.68	3.50	3.32	3.22	3.12	3.02	2.92	2.81	2.69
21	9.83	6.89	5.73	5.09	4.68	4.39	4.18	4.01	3.88	3.77	3.60	3.43	3.21	3.15	3.05	2.95	2.84	2.73	2.61
22	9.73	6.81	5.65	5.02	4.61	4.32	4.11	3.94	3.81	3.70	3.51	3.36	3.18	3.08	2.98	2.88	2.77	2.66	2.55
23	9.63	6.73	5.58	4.95	4.54	4.26	4.05	3.88	3.75	3.64	3.47	3.30	3.12	3.02	2.92	2.82	2.71	2.60	2.48
24	9.55	6.66	5.52	4.89	4.49	4.20	3.99	3.83	3.69	3.59	3.42	3.25	3.06	2.97	2.87	2.77	2.66	2.55	2.43
25	9.48	6.60	5.46	4.81	4.43	4.15	3.94	3.78	3.64	3.54	3.37	3.20	3.01	2.92	2.82	2.72	2.61	2.50	2.38
26	9.41	6.54	5.41	4.79	4.38	4.10	3.89	3.73	3.60	3.49	3.33	3.15	2.97	2.87	2.77	2.67	2.56	2.45	2.33
27	9.31	6.49	5.36	4.74	4.34	4.06	3.85	3.69	3.56	3.45	3.28	3.11	2.93	2.83	2.73	2.63	2.52	2.41	2.29
28	9.28	6.44	5.32	4.70	4.30	4.02	3.81	3.65	3.52	3.41	3.25	3.07	2.89	2.79	2.69	2.59	2.48	2.37	2.25
29	9.23	6.40	5.28	4.66	4.26	3.98	3.77	3.61	3.48	3.38	3.21	3.01	2.86	2.76	2.66	2.56	2.45	2.33	2.21
30	9.18	6.35	5.21	4.62	4.23	3.95	3.74	3.58	3.45	3.34	3.18	3.01	2.82	2.73	2.63	2.52	2.42	2.30	2.18
40	8.83	6.07	4.98	4.37	3.99	3.71	3.51	3.35	3.22	3.12	2.95	2.78	2.60	2.50	2.40	2.30	2.18	2.06	1.93
60	8.49	5.79	4.73	4.11	3.76	3.49	3.29	3.13	3.01	2.90	2.74	2.57	2.39	2.29	2.19	2.08	1.96	1.83	1.69
120	8.18	5.54	4.50	3.92	3.55	3.28	3.09	2.93	2.81	2.71	2.51	2.37	2.19	2.07	1.98	1.87	1.75	1.61	1.43
∞	7.88	5.30	4.28	3.72	3.35	3.07	2.90	2.71	2.62	2.52	2.36	2.19	2.00	1.90	1.79	1.67	1.53	1.36	1.00

- Compute the variable $F*$, defined as the ratio of the larger variance estimate to the smaller:

$$F* = \max(s_1^2/s_2^2, \ s_2^2/s_1^2)$$

- The analysis may be interpreted as follows:
  If $F* > F$, we may assert that $\sigma_1^2 \neq \sigma_2^2$ with $100\ (1-\alpha)$ confidence.
- If $F* < F$, we conclude that there is insufficient evidence to support the assertion that $\sigma_1^2 \neq \sigma_2^2$.

The following example illustrates the analysis procedure.

Illustration 1

A particular elastomer is produced in a stirred tank reactor with a solution catalyst. We wish to replace the catalyst with a less expensive one and still maintain good product properties (in particular, equivalent cure properties). Hence, we need to assess whether the product variability is the same for the product made from the different catalysts. The cure rates of seven samples, each from a respective catalyst, were measured and found to have standard deviations of 2.0 and 2.8, respectively.

The evaluation is as follows:

- Select $\alpha = 0.05$
- $s_1^2 = 4.0,\ \nu_1 = 6$
  $s_2^2 = 7.8,\ \nu_2 = 6$
- From Table 4-1,

$$F = 5.82 \qquad \text{for } \nu_N = 6 \text{ and } \nu_D = 6$$

- $F* = s_2^2/s_1^2$

  $= 7.84/4.0 = 1.96$

- Since $F* < F$ we do not have sufficient data to support that the variability in cure between the products is different.

If we assume the products are equivalent, then the best estimate of the variability is a pooled estimate of the standard deviation:

$$s_{pooled}^2 = (\nu^1 s_1^2 + \nu_2 s_2^2)/(\nu_1 + \nu_2)$$

$$s_{pooled}^2 = [6(4) + 6(7.84)]/12 = 5.92$$

$$s_{pooled} = 2.43 \text{ with } 11 \text{ degrees of freedom}$$

In the example given above we have calculated a pooled standard deviation by assuming $\sigma_1^2 = \sigma_2^2$. If this is true, then

$$S_{pooled} = \sqrt{\frac{\nu_1 s_1^2 + \nu_2 s_2^2}{\nu_1 + \nu_2}} \tag{3}$$

and the delta is:

$$s_\Delta = S_{pooled} \sqrt{1/n_1 + 1/n_2} \tag{4}$$

If in fact $\sigma_1^2 \neq \sigma_2^2$, then

$$s_\Delta = \sqrt{s_1^2/n_1 + s_2^2/n_2} \tag{5}$$

$$\nu_\Delta = \frac{(s_1^2/n_1 + s_2^2/n_2)^2}{\dfrac{(s_1^2/n_1)^2}{n_1 - 1} + \dfrac{(s_2^2/n_2)^2}{n_2 - 1}} \tag{6}$$

ν_Δ should be rounded off to the nearest integer.

One-Sided Comparison of Two Variances

This one-sided test is employed to control the risk in one direction only. Its basis is as follows:

$H_o : \sigma_1^2 \leq \sigma_2^2$, then we reject this hypothesis and assert that

$H_o : \sigma_1^2 > \sigma_2^2$ with α-risk.

The procedure for this method is as follows:

- Select the α-risk for H_o: $\sigma_1^2 \leq \sigma_2^2$.
- Compute s_1^2, s_2^2 (i.e., the variance of samples 1 and 2).
- Use Table 4-2 to obtain an F-value for $100(1-\alpha)\%$ confidence and ν_N, ν_D degrees of freedom.
- Compute F^* from the following ratio:

$$F^* = s_1^2/s_2^2$$

- Evaluate the results as follows:
  - If $F^* \geq F$, then we may assert that $\sigma_1^2 > \sigma_2^2$ with $100(1 - \alpha)\%$ confidence.
  - If $F^* < F$ we must conclude that there is insufficient evidence to support the assertation that $\sigma_1^2 > \sigma_2^2$.

ν_D \ ν_N	1	2	3	4	5	6	7	8	9	10	12	15	20	24	30	40	60	120	∞
1	39.86	49.50	53.59	55.83	57.24	58.20	58.91	59.44	59.86	60.19	60.71	61.22	61.74	62.00	62.26	62.53	62.79	63.06	63.33
2	8.53	9.00	9.16	9.24	9.29	9.33	9.35	9.37	9.38	9.39	9.41	9.42	9.44	9.45	9.46	9.47	9.47	9.48	9.49
3	5.54	5.46	5.39	5.34	5.31	5.28	5.27	5.25	5.24	5.23	5.22	5.20	5.18	5.18	5.17	5.16	5.15	5.14	5.13
4	4.54	4.32	4.19	4.11	4.05	4.01	3.98	3.95	3.94	3.92	3.90	3.87	3.84	3.83	3.82	3.80	3.79	3.78	3.76
5	4.06	3.78	3.62	3.52	3.45	3.40	3.37	3.34	3.32	3.30	3.27	3.24	3.21	3.19	3.17	3.16	3.14	3.12	3.10
6	3.78	3.46	3.29	3.18	3.11	3.05	3.01	2.98	2.96	2.94	2.90	2.87	2.84	2.82	2.80	2.78	2.76	2.74	2.72
7	3.59	3.26	3.07	2.96	2.88	2.83	2.78	2.75	2.72	2.70	2.67	2.63	2.59	2.58	2.56	2.54	2.51	2.49	2.47
8	3.46	3.11	2.92	2.81	2.73	2.67	2.62	2.59	2.56	2.54	2.50	2.46	2.42	2.40	2.38	2.36	2.34	2.32	2.29
9	3.36	3.01	2.81	2.69	2.61	2.55	2.51	2.47	2.44	2.42	2.38	2.34	2.30	2.28	2.25	2.23	2.21	2.18	2.16
10	3.29	2.92	2.73	2.61	2.52	2.46	2.41	2.38	2.35	2.32	2.28	2.24	2.20	2.18	2.16	2.13	2.11	2.08	2.06
11	3.23	2.86	2.66	2.54	2.45	2.39	2.34	2.30	2.27	2.25	2.21	2.17	2.12	2.10	2.08	2.05	2.03	2.00	1.97
12	3.18	2.81	2.61	2.48	2.39	2.33	2.28	2.24	2.21	2.19	2.15	2.10	2.06	2.04	2.01	1.99	1.96	1.93	1.90
13	3.14	2.76	2.56	2.43	2.35	2.28	2.23	2.20	2.16	2.14	2.10	2.05	2.01	1.98	1.96	1.93	1.90	1.88	1.85
14	3.10	2.73	2.52	2.39	2.31	2.24	2.19	2.15	2.12	2.10	2.05	2.01	1.96	1.94	1.91	1.89	1.86	1.83	1.80
15	3.07	2.70	2.49	2.36	2.27	2.21	2.16	2.12	2.09	2.06	2.02	1.97	1.92	1.90	1.87	1.85	1.82	1.79	1.76
16	3.05	2.67	2.46	2.33	2.24	2.18	2.13	2.09	2.06	2.03	1.99	1.94	1.89	1.87	1.84	1.81	1.78	1.75	1.72
17	3.03	2.64	2.44	2.31	2.22	2.15	2.10	2.06	2.03	2.00	1.96	1.91	1.86	1.84	1.81	1.78	1.75	1.72	1.69
18	3.01	2.62	2.42	2.29	2.20	2.13	2.08	2.04	2.00	1.98	1.93	1.89	1.84	1.81	1.78	1.75	1.72	1.69	1.66
19	2.99	2.61	2.40	2.27	2.18	2.11	2.06	2.02	1.98	1.96	1.91	1.86	1.81	1.79	1.76	1.73	1.70	1.67	1.63
20	2.97	2.59	2.38	2.25	2.16	2.09	2.04	2.00	1.96	1.94	1.89	1.84	1.79	1.77	1.74	1.71	1.68	1.64	1.61
21	2.96	2.57	2.36	2.23	2.14	2.08	2.02	1.98	1.95	1.92	1.87	1.83	1.78	1.75	1.72	1.69	1.66	1.62	1.59
22	2.95	2.56	2.35	2.22	2.13	2.06	2.01	1.97	1.93	1.90	1.86	1.81	1.76	1.73	1.70	1.67	1.64	1.60	1.57
23	2.94	2.55	2.34	2.21	2.11	2.05	1.99	1.95	1.92	1.89	1.84	1.80	1.74	1.72	1.69	1.66	1.62	1.59	1.55
24	2.93	2.54	2.33	2.19	2.10	2.04	1.98	1.94	1.91	1.88	1.83	1.78	1.73	1.70	1.67	1.64	1.61	1.57	1.53
25	2.92	2.53	2.32	2.18	2.09	2.02	1.97	1.93	1.89	1.87	1.82	1.77	1.72	1.69	1.66	1.63	1.59	1.56	1.52
26	2.91	2.52	2.31	2.17	2.08	2.01	1.96	1.92	1.88	1.86	1.81	1.76	1.71	1.68	1.65	1.61	1.58	1.54	1.50
27	2.90	2.51	2.30	2.17	2.07	2.00	1.95	1.91	1.87	1.85	1.80	1.75	1.70	1.67	1.64	1.60	1.57	1.53	1.49
28	2.89	2.50	2.29	2.16	2.06	2.00	1.94	1.90	1.87	1.84	1.79	1.74	1.69	1.66	1.63	1.59	1.56	1.52	1.48
29	2.89	2.50	2.28	2.15	2.06	1.99	1.93	1.89	1.86	1.83	1.78	1.73	1.68	1.65	1.62	1.58	1.55	1.51	1.47
30	2.88	2.49	2.28	2.14	2.05	1.98	1.93	1.88	1.85	1.82	1.77	1.72	1.67	1.64	1.61	1.57	1.54	1.50	1.46
40	2.84	2.41	2.23	2.09	2.00	1.93	1.87	1.83	1.79	1.76	1.71	1.66	1.61	1.57	1.54	1.51	1.47	1.42	1.38
60	2.79	2.39	2.18	2.04	1.95	1.87	1.82	1.77	1.74	1.71	1.66	1.60	1.54	1.51	1.48	1.44	1.40	1.35	1.29
120	2.75	2.35	2.13	1.99	1.90	1.82	1.77	1.72	1.68	1.65	1.60	1.55	1.48	1.45	1.41	1.37	1.32	1.26	1.19
∞	2.71	2.30	2.08	1.94	1.85	1.77	1.72	1.67	1.63	1.60	1.55	1.49	1.42	1.38	1.34	1.30	1.24	1.17	1.00

(continued)

Table 4-2. (continued).

α = 0.05

ν_D \ ν_N	1	2	3	4	5	6	7	8	9	10	12	15	20	24	30	40	60	120	∞
1	161.4	199.5	215.7	224.6	230.2	234.0	236.8	238.9	240.5	241.9	243.9	245.9	248.0	249.1	250.1	251.1	252.2	253.3	254.3
2	18.51	19.00	19.16	19.25	19.30	19.33	19.35	19.37	19.38	19.40	19.41	19.43	19.45	19.46	19.46	19.47	19.48	19.49	19.50
3	10.13	9.55	9.28	9.12	9.01	8.94	8.89	8.85	8.81	8.79	8.74	8.70	8.66	8.64	8.62	8.59	8.57	8.55	8.53
4	7.71	6.94	6.59	6.39	6.26	6.16	6.09	6.04	6.00	5.96	5.91	5.86	5.80	5.77	5.75	5.72	5.69	5.66	5.63
5	6.61	5.79	5.41	5.19	5.05	4.95	4.88	4.82	4.77	4.74	4.68	4.62	4.56	4.53	4.50	4.46	4.43	4.40	4.36
6	5.99	5.14	4.76	4.53	4.39	4.28	4.21	4.15	4.10	4.06	4.00	3.94	3.87	3.84	3.81	3.77	3.74	3.70	3.67
7	5.59	4.74	4.35	4.12	3.97	3.87	3.79	3.73	3.68	3.64	3.57	3.51	3.44	3.41	3.38	3.34	3.30	3.27	3.23
8	5.32	4.46	4.07	3.84	3.69	3.58	3.50	3.44	3.39	3.35	3.28	3.22	3.15	3.12	3.08	3.04	3.01	2.97	2.93
9	5.12	4.26	3.86	3.63	3.48	3.37	3.29	3.23	3.18	3.14	3.07	3.01	2.94	2.90	2.86	2.83	2.79	2.75	2.71
10	4.96	4.10	3.71	3.48	3.33	3.22	3.14	3.07	3.02	2.98	2.91	2.85	2.77	2.74	2.70	2.66	2.62	2.58	2.54
11	4.84	3.98	3.59	3.36	3.20	3.09	3.01	2.95	2.90	2.85	2.79	2.72	2.65	2.61	2.57	2.53	2.49	2.45	2.40
12	4.75	3.89	3.49	3.26	3.11	3.00	2.91	2.85	2.80	2.75	2.69	2.62	2.54	2.51	2.47	2.43	2.38	2.34	2.30
13	4.67	3.81	3.41	3.18	3.03	2.92	2.83	2.77	2.71	2.67	2.60	2.53	2.46	2.42	2.38	2.34	2.30	2.25	2.21
14	4.60	3.74	3.34	3.11	2.96	2.85	2.76	2.70	2.65	2.60	2.53	2.46	2.39	2.35	2.31	2.27	2.22	2.18	2.13
15	4.54	3.68	3.29	3.06	2.90	2.79	2.71	2.64	2.59	2.54	2.48	2.40	2.33	2.29	2.25	2.20	2.16	2.11	2.07
16	4.49	3.63	3.24	3.01	2.85	2.74	2.66	2.59	2.54	2.49	2.42	2.35	2.28	2.24	2.19	2.15	2.11	2.06	2.01
17	4.45	3.59	3.20	2.96	2.81	2.70	2.61	2.55	2.49	2.45	2.38	2.31	2.23	2.19	2.15	2.10	2.06	2.01	1.96
18	4.41	3.55	3.16	2.93	2.77	2.66	2.58	2.51	2.46	2.41	2.34	2.27	2.19	2.15	2.11	2.06	2.02	1.97	1.92
19	4.38	3.52	3.13	2.90	2.74	2.63	2.54	2.48	2.42	2.38	2.31	2.23	2.16	2.11	2.07	2.03	1.98	1.93	1.88
20	4.35	3.49	3.10	2.87	2.71	2.60	2.51	2.45	2.39	2.35	2.28	2.20	2.12	2.08	2.04	1.99	1.95	1.90	1.84
21	4.32	3.47	3.07	2.84	2.68	2.57	2.49	2.42	2.37	2.32	2.25	2.18	2.10	2.05	2.01	1.96	1.92	1.87	1.81
22	4.30	3.44	3.05	2.82	2.66	2.55	2.46	2.40	2.34	2.30	2.23	2.15	2.07	2.03	1.98	1.94	1.89	1.84	1.78
23	4.28	3.42	3.03	2.80	2.64	2.53	2.44	2.37	2.32	2.27	2.20	2.13	2.05	2.01	1.96	1.91	1.86	1.81	1.76
24	4.26	3.40	3.01	2.78	2.62	2.51	2.42	2.36	2.30	2.25	2.18	2.11	2.03	1.98	1.94	1.89	1.84	1.79	1.73
25	4.24	3.39	2.99	2.76	2.60	2.49	2.40	2.34	2.28	2.24	2.16	2.09	2.01	1.96	1.92	1.87	1.82	1.77	1.71
26	4.23	3.37	2.98	2.74	2.59	2.47	2.39	2.32	2.27	2.22	2.15	2.07	1.99	1.95	1.90	1.85	1.80	1.75	1.69
27	4.21	3.35	2.96	2.73	2.57	2.46	2.37	2.31	2.25	2.20	2.13	2.06	1.97	1.93	1.88	1.84	1.79	1.73	1.67
28	4.20	3.34	2.95	2.71	2.56	2.45	2.36	2.29	2.24	2.19	2.12	2.04	1.96	1.91	1.87	1.82	1.77	1.71	1.65
29	4.18	3.33	2.93	2.70	2.55	2.43	2.35	2.28	2.22	2.18	2.10	2.03	1.94	1.90	1.85	1.81	1.75	1.70	1.64
30	4.17	3.32	2.92	2.69	2.53	2.42	2.33	2.27	2.21	2.16	2.09	2.01	1.93	1.89	1.84	1.79	1.74	1.68	1.62
40	4.08	3.23	2.84	2.61	2.45	2.34	2.25	2.18	2.12	2.08	2.00	1.92	1.84	1.79	1.74	1.69	1.61	1.58	1.51
60	4.00	3.15	2.76	2.53	2.37	2.25	2.17	2.10	2.04	1.99	1.92	1.84	1.75	1.70	1.65	1.59	1.53	1.47	1.39
120	3.92	3.07	2.68	2.45	2.29	2.17	2.09	2.02	1.96	1.91	1.83	1.75	1.66	1.61	1.55	1.50	1.43	1.35	1.25
∞	3.84	3.00	2.60	2.37	2.21	2.10	2.01	1.94	1.88	1.83	1.75	1.67	1.57	1.52	1.46	1.39	1.32	1.22	1.00

80

Table 4-2. (continued).

α = 0.01

ν_D \\ ν_N	1	2	3	4	5	6	7	8	9	10	12	15	20	24	30	40	60	120	∞
1	4052	4999.5	5403	5625	5761	5859	5928	5982	6022	6056	6106	6157	6209	6235	6261	6287	6313	6339	6366
2	98.50	99.00	99.17	99.25	99.30	99.33	99.36	99.37	99.39	99.40	99.42	99.43	99.45	99.46	99.47	99.47	99.48	99.49	99.50
3	34.12	30.82	29.46	28.71	28.24	27.91	27.67	27.49	27.35	27.23	27.05	26.87	26.69	26.60	26.50	26.41	26.32	26.22	26.13
4	21.20	18.00	16.69	15.98	15.52	15.21	14.98	14.80	14.66	14.55	14.37	14.20	14.02	13.93	13.84	13.75	13.65	13.56	13.46
5	16.26	13.27	12.06	11.39	10.97	10.67	10.46	10.29	10.16	10.05	9.89	9.72	9.55	9.47	9.38	9.29	9.20	9.11	9.02
6	13.75	10.92	9.78	9.15	8.75	8.47	8.26	8.10	7.98	7.87	7.72	7.56	7.40	7.31	7.23	7.14	7.06	6.97	6.88
7	12.25	9.55	8.45	7.85	7.46	7.19	6.99	6.84	6.72	6.62	6.47	6.31	6.16	6.07	5.99	5.91	5.82	5.74	5.65
8	11.26	8.65	7.59	7.01	6.63	6.37	6.18	6.03	5.91	5.81	5.67	5.52	5.36	5.28	5.20	5.12	5.03	4.95	4.86
9	10.56	8.02	6.99	6.42	6.06	5.80	5.61	5.47	5.35	5.26	5.11	4.96	4.81	4.73	4.65	4.57	4.48	4.40	4.31
10	10.04	7.56	6.55	5.99	5.64	5.39	5.20	5.06	4.94	4.85	4.71	4.56	4.41	4.33	4.25	4.17	4.08	4.00	3.91
11	9.65	7.21	6.22	5.67	5.32	5.07	4.89	4.74	4.63	4.54	4.40	4.25	4.10	4.02	3.94	3.86	3.78	3.69	3.60
12	9.33	6.93	5.95	5.41	5.06	4.82	4.64	4.50	4.39	4.30	4.16	4.01	3.86	3.78	3.70	3.62	3.54	3.45	3.36
13	9.07	6.70	5.74	5.21	4.86	4.62	4.44	4.30	4.19	4.10	3.96	3.82	3.66	3.59	3.51	3.43	3.34	3.25	3.17
14	8.86	6.51	5.56	5.04	4.69	4.46	4.28	4.14	4.03	3.94	3.80	3.66	3.51	3.43	3.35	3.27	3.18	3.09	3.00
15	8.68	6.36	5.42	4.89	4.56	4.32	4.14	4.00	3.89	3.80	3.67	3.52	3.37	3.29	3.21	3.13	3.05	2.96	2.87
16	8.53	6.23	5.29	4.77	4.44	4.20	4.03	3.89	3.78	3.69	3.55	3.41	3.26	3.18	3.10	3.02	2.93	2.84	2.75
17	8.40	6.11	5.18	4.67	4.34	4.10	3.93	3.79	3.68	3.59	3.46	3.31	3.16	3.08	3.00	2.92	2.83	2.75	2.65
18	8.29	6.01	5.09	4.58	4.25	4.01	3.84	3.71	3.60	3.51	3.37	3.23	3.08	3.00	2.92	2.81	2.75	2.66	2.57
19	8.18	5.93	5.01	4.50	4.17	3.94	3.77	3.63	3.52	3.43	3.30	3.15	3.00	2.92	2.84	2.76	2.67	2.58	2.49
20	8.10	5.85	4.94	4.43	4.10	3.87	3.70	3.56	3.46	3.37	3.23	3.09	2.94	2.86	2.78	2.69	2.61	2.52	2.42
21	8.02	5.78	4.87	4.37	4.04	3.81	3.64	3.51	3.40	3.31	3.17	3.03	2.88	2.80	2.72	2.64	2.55	2.46	2.36
22	7.95	5.72	4.82	4.31	3.99	3.76	3.59	3.45	3.35	3.26	3.12	2.98	2.83	2.75	2.67	2.58	2.50	2.40	2.31
23	7.88	5.66	4.76	4.26	3.94	3.71	3.54	3.41	3.30	3.21	3.07	2.93	2.78	2.70	2.62	2.54	2.45	2.35	2.26
24	7.82	5.61	4.72	4.22	3.90	3.67	3.50	3.36	3.26	3.17	3.03	2.89	2.74	2.66	2.58	2.49	2.40	2.31	2.21
25	7.77	5.57	4.68	4.18	3.85	3.63	3.46	3.32	3.22	3.13	2.99	2.85	2.70	2.62	2.54	2.45	2.36	2.27	2.17
26	7.72	5.53	4.64	4.14	3.82	3.59	3.42	3.29	3.18	3.09	2.96	2.81	2.66	2.58	2.50	2.42	2.33	2.23	2.13
27	7.68	5.49	4.60	4.11	3.78	3.56	3.39	3.26	3.15	3.06	2.93	2.78	2.63	2.55	2.47	2.38	2.29	2.20	2.10
28	7.64	5.45	4.57	4.07	3.75	3.53	3.36	3.23	3.12	3.03	2.90	2.75	2.60	2.52	2.44	2.35	2.26	2.17	2.06
29	7.60	5.42	4.54	4.04	3.73	3.50	3.33	3.20	3.09	3.00	2.87	2.73	2.57	2.49	2.41	2.33	2.23	2.14	2.03
30	7.56	5.39	4.51	4.02	3.70	3.47	3.30	3.17	3.07	2.98	2.84	2.70	2.55	2.47	2.39	2.30	2.21	2.11	2.01
40	7.31	5.18	4.31	3.83	3.51	3.29	3.12	2.99	2.89	2.80	2.66	2.52	2.37	2.29	2.20	2.11	2.02	1.92	1.80
60	7.08	4.98	4.13	3.65	3.34	3.12	2.95	2.82	2.72	2.63	2.50	2.35	2.20	2.12	2.03	1.94	1.84	1.73	1.60
120	6.85	4.79	3.95	3.48	3.17	2.96	2.79	2.66	2.56	2.47	2.34	2.19	2.03	1.95	1.86	1.76	1.66	1.53	1.38
∞	6.63	4.61	3.78	3.32	3.02	2.80	2.64	2.51	2.41	2.32	2.18	2.04	1.88	1.79	1.70	1.59	1.47	1.32	1.00

81

Illustration 2

Two ASTM methods for measuring the ethylene content of high crystallinity polymers are being considered. Method 1 has a standard deviation of $\sigma_1 = 0.15$ (based on 300 measurements), and Method 2 has $\sigma_2 = 0.077$ (based on 5 measurements). Which method is better?

- We choose $\alpha = 0.01$
- Compute the standard deviations squared

$s_1^2 = (0.15)^2 = 0.0225$ ($\nu_1 = 299$, for practicality we may assume $\nu_1 = \infty$).

$s_2^2 = (0.077)^2 = 0.0059$ ($\nu_2 = 4$)

$F^* = 0.0225/0.0059 = 3.81$

- From Table 4-2; $\nu_N = \infty$ and $\nu_D = 4$, F = 13.46.

Since $F^* < F$, we are not 99% confident that $\sigma_1^2 > \sigma_2^2$. Hence, more data is required to make an assessment of the methods.

A very practical consideration is the number of measurements needed on each of two populations in order to assess a given ratio θ, of the two population variances. This problem can be tackled using the following procedure and Table 4-3. Table 4-3 gives the number of observations required for the comparison of two population variances using the F-test. The table provides values of the ratio of the two population variances, σ_1^2/σ_2^2 which remains undetected with a probability β if the variance ratio test (F-test) is used at a significance level of α on the ratio of the estimates of the two variances (s_1^2/s_2^2), each being based on the same number of degrees of freedom. The recommended procedure is as follows:

- Select the risk of asserting that there is a difference when none exists (α), and the risk of failing (β) to detect a difference when a ratio as large as θ exists.
- Decide whether to apply a two-sided or a one-sided comparison after the measurements are made.
- Use Table 4-3 to obtain n as follows:
  - Find the appropriate column corresponding to the specified α and β values.
  - Obtain the θ ratio from that column.
  - Reading to the left from the θ value, find the required number of degrees of freedom, interpolating for that value if necessary. The n value obtained represents the required number of measurements on each population, and the number of degrees of freedom will be $\nu = n-1$.

Table 4-3. Number of Required Observations for Comparison of Two Population Variances Based on F-Test. *

ν	Double-Sided $\alpha = 0.02$ Single-Sided $\alpha = 0.01$				$\alpha = 0.10$ $\alpha = 0.05$			
	$\beta = 0.01$	$\beta = 0.05$	$\beta = 0.1$	$\beta = 0.5$	$\beta = 0.01$	$\beta = 0.05$	$\beta = 0.1$	$\beta = 0.5$
1	16,420,000	654,200	161,500	4052	654,200	26,070	6,436	161.5
2	9,000	1,881	891.0	99.00	1,881	361.0	171.0	19.00
3	867.7	273.3	158.8	29.46	273.3	86.06	50.01	9.277
4	255.3	102.1	65.62	15.98	102.1	40.81	26.24	6.388
5	120.3	55.39	37.87	10.97	55.39	25.51	17.44	5.050
6	71.67	36.27	25.86	8.466	38.27	18.35	13.09	4.284
7	48.90	26.48	19.47	6.993	26.48	14.34	10.55	3.787
8	36.35	20.73	15.61	6.029	20.73	11.82	8.902	3.438
9	28.63	17.01	13.06	5.351	17.01	10.11	7.757	3.179
10	23.51	14.44	11.26	4.849	14.44	8.870	6.917	2.978
12	17.27	11.16	8.923	4.155	11.16	7.218	5.769	2.687
15	12.41	8.466	6.946	3.522	8.466	5.777	4.740	2.404
20	8.630	6.240	5.270	2.938	6.240	4.512	3.810	2.124
24	7.071	5.275	4.526	2.659	5.275	3.935	3.376	1.984
30	5.693	4.392	3.833	2.386	4.392	3.389	2.957	1.841
40	4.470	3.579	3.183	2.114	3.579	2.866	2.549	1.693
60	3.372	2.817	2.562	1.836	2.817	2.354	2.141	1.534
120	2.350	2.072	1.939	1.533	2.072	1.828	1.710	1.352
∞	1.000	1.000	1.000	1.000	1.000	1.000	1.000	1.000

*The body of this table gives the value of the ratio of the two population variances, σ_1^2/σ_2^2 which remains undetected with probability β if a variance ratio test (i.e. F test) is used at a significance level of α on the ratio of the estimates of the two variances, s_1^2/s_2^2, each being based on the same number of degrees of freedom (ν).

Illustration 3

A rather lengthy test is being used to detect gels in the production of polymers. An alternate procedure which is shorter and less manpower-intensive is being considered. We decide, however, that if the existing test method has a standard deviation of more than half of that of the new procedure, we will stay with the more intensive method. How many measurements are needed to assess this?

- Select $\alpha = 0.05$ and $\beta = 0.05$.
- A one-sided comparison will be applied.
- The variance ratio that we do not want to miss is:

  $\theta = (1)^2/(1/2)^2 = 4$

  From Table 4-3, $\nu \cong 24$.

Hence, we will need at least 25 measurements.

Comparison of Means

We now pose the problem as to how many measurements are needed on each of two means in order to compare them to be able to detect a difference. That is, $\delta = \mu_1 - \mu_2$. The following procedure can be used:

- Select α and β. That is, we choose the risk of asserting that there is a difference when none exists, and the risk of failing (β) to detect a difference when one of δ exists.
- We decide on whether a one-sided or a two-sided comparison will be made after the data is collected.
- Obtain an estimate of σ either from historical data or best guess. We assume in this analysis that both populations have the same σ-values.
- Use Table 2-7 in Chapter 2 to obtain the value of n—the number of measurements required for the appropriate α, β and δ/σ values.

Illustration 4

The coke gas yields (%) from a dilute phase riser reactor (DRR) and a conventional flexicoker (FC) are as follows:

DRR 86, 94, 88, 90, 89, 92, 82, 89, 78, 78

FC 81, 85, 83, 90, 92, 88, 86, 84, 78, 73

1 Does one of these reactors produce higher yields?

2 Determine how many measurements on each reactor would be needed to reveal a 2.5% yield increase or better.

For Part 1:

- Choose $\alpha = 0.05$.

- $\bar{x}_1 = 86.6; \; s_1^2 = 30.9; \; \nu_1 = 9$
  $\bar{x}_2 = 84.0; \; s_2^2 = 32.0; \; \nu_2 = 9$
- From Equation (5) $s_\Delta = 2.51$
  From Equation (6) $\nu_\Delta = 18$
  $\Delta = \bar{x}_1 - \bar{x}_2 = 2.6$
- From Table 2-3 in Chapter 2, $t = 1.734$
- $t^* = \Delta / s_\Delta$
  $t^* = 2.6/2.51 = 1.04$
- Since $t^* < t$, there is not enough evidence to be able to say that one reactor is better than the other.

In this part of the problem we have applied a one-sided comparison of two means. If $t^* > t$, then we could assert that $\mu_1 > \mu_2$ with 95% confidence.

For Part 2:

- Choose $\alpha = 0.05$ and $\beta = 0.10$.
- We decide that a one-sided test will be performed after data collection.
- From Equation (3), $s_{pooled} = 5.61\%$.
- From Table 2-7 in Chapter 2, for $\delta/\sigma = 2.5/\sqrt{30.9} = 0.45$:
  $n = 86$ (i.e., 86 measurements for each reactor are required).

We now direct attention to the problem of making a two-sided comparison of two means. The means of two populations can be compared in a statistical test after the issue of equality of variances is settled using Equations (3) through (6). The two-sided test for equality of two means is outlined by the procedure given below. In this analysis, the following hypothesis is tested:

$$H_o': \mu_1 = \mu_2 \text{ or } \mu_1 - \mu_2 = 0.$$

- Select the α-risk value.
- Compute the arithmetic means ($\bar{x}_1, \bar{x}_2$) and the standard deviations (s_1, s_2).
- Compute the Δ value: $\Delta = \bar{x} - \bar{x}'$
  Compute the standard deviation of Δ, s_Δ (Equation (4)).
- Use Table 2-2 from Chapter 2 to obtain a t-value for $100(1 - \alpha)\%$ confidence and ν_Δ degrees of freedom.
- Compute $t^* = |\Delta|/s_\Delta$.
- Interpret the results in the following manner:
  —If $t^* > t$, we may assert that $\mu_1 \neq \mu_2$ with $100 (1 - \alpha)\%$ confidence.
  —If $t^* < t$, we conclude that there is not sufficient evidence to support that μ_1 and μ_2 differ.

Illustration 5

An oil extended (O/E) polymer is produced for a variety of applications. One of our steady customers uses the product as a blend with a non-O/E grade, but finds he must add another 10 parts of oil during compounding in order to achieve a good mix. On one shipment the customer complains that we shipped him off-spec product and that their compound Mooneys are too high for producing their extruded product. The plant manufacturing data, however, indicates the product to be within spec. One possible scenario is that the customer neglected to add the additional 10 parts of oil during their compound mix. To test this hypothesis, we prepare two mixes using the customer's formulation: Mix A—normal compound, and Mix B—withholding 10 parts oil. The following data are the compound Mooney viscosities measured from several samples of that particular production run:

Mix A: 42   43   40   47   45   43.5

Mix B: 41   47   49   50   46   48   49   50

Can we conclude that the customer did not prepare the recipe properly or must we suspect a problem with our polymer?

- We choose $\alpha = 0.05$.
- $\bar{x}_1 = 43.4$; $s_1^2 = 5.84$; $n_1 = 6$
  $\bar{x}_2 = 47.5$; $s_2^2 = 8.86$; $n_2 = 8$
- $\Delta = 4.10$
    From Equation (5), $s_\Delta = \sqrt{5.84/6 + 8.86/8} = 1.44$
    From Equation (6) $\nu_\Delta = 11.87 \cong 12$.
- From Table 2-2 in Chapter 2, $t = 2.179$.
- $t^* = |4.10|/1.44 = 2.85$
- Since $t^* > t$ we have reason to suspect that the customer did not properly prepare the mix.

A final shortcut technique, applicable to comparing two means, addresses the problem of extraneous errors. That is, extraneous factors during an experiment may introduce large contributions to error. Hence, in the sampling of two populations, σ values may become inflated. This makes the test of $\mu_1 = \mu_2$ less sensitive by obscuring small differences. Some of this problem can be eliminated through the method of *pairing observations*. When populations are generated in parallel, we can pair each of the n samples of the first population with a corresponding sample from the second population and apply the following analysis method:

- First choose the risk (α) of asserting that there is a difference when none exists.
- Obtain a value of t for $100(1 - \alpha)\%$ confidence and n - 1 degrees of

freedom from Table 2-2, Chapter 2 (for a two-sided test) or from Table 2-3, Chapter 2 (for a one-sided test).

- Set up a table and compute d_i, the difference in pairs ($= x_{1i} - x_{2i}$); the mean difference $\bar{d} = d_i/n$, $\Sigma i = 1, \ldots, n$ and the deviation of the differences:

$$s_d^2 = \Sigma(d_1 - \bar{d})^2/(n - 1) = \{\Sigma d_i^2) - (\Sigma d_i)^2/n\}/(n - 1)$$

- Compute t^*:

  For a two-sided test:

  $$t^* = |\bar{d}|/\sqrt{s_d^2/n}$$

  For a one-sided test:

  $$t^* = \bar{d}/\sqrt{s_d^2/n}$$

- Base conclusions on the following criteria:
  - If $t^* > t$, we may assume that $\mu_1 \neq \mu_2$ (or $\mu_1 > \mu_2$ for a one-sided test) with $100(1 - \alpha)\%$ confidence.
  - If $t^* < t$, we must conclude that there is insufficient evidence to support that $\mu_1 \neq \mu_2$ (or $\mu_1 > \mu_2$ for a one-sided test).

Illustration 6

We compare the collapse ratio of an experimental polymer with an existing grade used in the manufacture of automotive hose. One concern in the study is that the variability of extrusion and the condition of the hose die may introduce large variability to the data. We therefore run the polymer samples under identical extrusion conditions (i.e., same head pressure, screw rpm, extruder wall temperatures) and make a side-by-side comparison of each data point. The data are:

Standard Polymer:	0.78	0.79	0.77	0.76	0.74	0.77
Prototype Polymer:	0.81	0.73	0.68	0.93	0.79	0.78

The analysis is as follows:

- Choose $\alpha = 0.05$
- $\bar{x}_1 = 0.77$; $s_1^2 = 2.97 \times 10^{-4}$; $\nu_1 = 5$
  $\bar{x}_2 = 0.79$; $s_1^2 = 7.15 \times 10^{-3}$; $\nu_2 = 5$
- From Table 2-3, Chapter 2 (one-sided test), $t = 2.015$.
- $d_i = -0.03, 0.06, 0.09, -0.17, -0.05, -0.01$
  $\bar{d} = -0.02$
  $s_d^2 = 8.42 \times 10^{-3}$

- Now compute t^*

$$t^* = |-0.02|/\sqrt{8.42 \times 10^{-3}/6} = 0.53$$

Since $t^* < t$ we must conclude that there is not enough evidence to believe that the prototype is any better.

ANALYSIS OF VARIANCE

General Terminology

Before illustrating the technique of analysis of variance, some general concepts and definitions need to be introduced. Conceptually, we may view any piece of equipment or system as a "black box." The details of the box are often unimportant in understanding what factors or variables produce a response. In fact, it is these terms we wish to define. Figure 4-1 illustrates some basic concepts in the design of an experiment. The term *response* refers to the observed outcome of the experiment. For example, in the case of a chemical reactor, one of the key responses would most certainly be the product yield. The term *factor* refers to the variable imposed on the system by the experimenter. In the case of a reactor the factor or factors might be temperature, pressure, feed rates, poisons, etc.

The responses of an experimental unit may be classified according to scale:

- Quantitative or continuous (e.g., cure properties of a polymer)—these are generally the easiest to analyze.
- Qualitative or categorical (e.g., color or clarity of product or wastestreams)—these should be converted to some convenient semi-quantitative scale.
- Quantal or binary (e.g., population fatalities in evaluating lethal dosages of toxic materials)—this often requires several experiments to obtain a semi-continuous response.

Factors affecting the experiment may also be classified according to scale. For example, they may be classified as quantitative or continuous (e.g., pressure, temperature, residence time), or qualitative or categorical (e.g., solvent, catalyst type, baffled vs. non-baffled reactor).

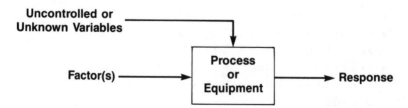

FIGURE 4-1. Basic concepts in the design of single factor experiments.

The factor may be applied to different degrees, hence, the term *factor level*. This refers to the actual setting or value of the factor during a given run within the experiment (e.g., 70°C for temperature or the use of silica catalyst A).

In designing an experiment, one should specify a series of runs with the objectives of studying the nature of the system's response to a particular factor and the ultimate effect, i.e., the change in response between two factor levels. Three elements to include in a well-designed factor experiment are:

- Specification of the factor and expected response.
- Specification of the factor levels.
- The number of replicates required for each factor level.

As an example, consider the experiment to assess the yield of syngas in a pressurized gasification process. The experiment might be:

	Pressure	
100 psia	250 psia	500 psia
X	X	X
X	X	X
X	X	X

In this case, we request 3 replicates per pressure level to test the response of yield to pressure.

To minimize the contribution of uncontrolled variables in the experiment, it's important that the experimental unit be allocated to factor levels in a randomized manner. Both the experimental order and the testing order should be randomized in order to minimize the effects of systematic errors in data taking. A completely randomized experimental design is one where the experimental units are indistinguishable and are randomly assigned to factor levels.

Each factor level x_i can generate a population of responses Y_i with mean μ_i and variance σ_i^2. The investigator's objective is to determine whether there is an effect, which can be stated in terms of the null hypothesis:

$$H_o : \mu_1 = \mu_2 = \ldots = \mu_k, \text{ for } k \text{ factor levels.}$$

If any of the means are not equal, then the hypothesis will be rejected.

Homogeneity of Variance

Earlier we illustrated the problem of testing the equality of two means. The basis for this analysis was homogeneity of variances (i.e., we assumed that the population variances were equal). This same assumption is applied in analysis of variance (ANOVA). The test used to assess the assumption of equal variances is called Cochran's test. If variances are found not to be equivalent,

Table 4-4. Values of Cochran's Test for Constant Variance.

α = .05

ν \ k	1	2	3	4	5	6	7	8	9	10	16	36	144	∞
2	0.9985	0.9750	0.9392	0.9057	0.8772	0.8534	0.8332	0.8159	0.8010	0.7880	0.7341	0.6602	0.5813	0.5000
3	0.9669	0.8709	0.7977	0.7457	0.7071	0.6771	0.6530	0.6333	0.6167	0.6025	0.5466	0.4748	0.4031	0.3333
4	0.9065	0.7679	0.6841	0.6287	0.5895	0.5598	0.5365	0.5175	0.5017	0.4884	0.4366	0.3720	0.3093	0.2500
5	0.8412	0.6838	0.5981	0.5441	0.5065	0.4783	0.4564	0.4387	0.4241	0.4118	0.3645	0.3066	0.2513	0.2000
6	0.7808	0.6161	0.5321	0.4803	0.4447	0.4184	0.3980	0.3817	0.3682	0.3568	0.3135	0.2612	0.2119	0.1667
7	0.7271	0.5612	0.4800	0.4307	0.3974	0.3726	0.3535	0.3384	0.3259	0.3154	0.2756	0.2278	0.1833	0.1429
8	0.6798	0.5157	0.4377	0.3910	0.3595	0.3362	0.3185	0.3043	0.2926	0.2829	0.2462	0.2022	0.1616	0.1250
9	0.6385	0.4775	0.4027	0.3584	0.3286	0.3067	0.2901	0.2768	0.2659	0.2568	0.2226	0.1820	0.1446	0.1111
10	0.6020	0.4450	0.3733	0.3311	0.3029	0.2823	0.2666	0.2541	0.2439	0.2353	0.2032	0.1655	0.1308	0.1000
12	0.5410	0.3924	0.3264	0.2880	0.2624	0.2439	0.2299	0.2187	0.2098	0.2020	0.1737	0.1403	0.1100	0.0833
15	0.4709	0.3346	0.2758	0.2419	0.2195	0.2034	0.1911	0.1815	0.1736	0.1671	0.1429	0.1144	0.0889	0.0667
20	0.3894	0.2705	0.2205	0.1921	0.1735	0.1602	0.1501	0.1422	0.1357	0.1303	0.1108	0.0879	0.0675	0.0500
24	0.3434	0.2354	0.1907	0.1656	0.1493	0.1374	0.1286	0.1216	0.1160	0.1113	0.0942	0.0743	0.0567	0.0417
30	0.2929	0.1980	0.1593	0.1377	0.1237	0.1137	0.1061	0.1002	0.0958	0.0921	0.0771	0.0604	0.0457	0.0333
40	0.2370	0.1576	0.1259	0.1082	0.0968	0.0887	0.0827	0.0780	0.0745	0.0713	0.0595	0.0462	0.0347	0.0250
60	0.1737	0.1131	0.0895	0.0765	0.0682	0.0623	0.0583	0.0552	0.0520	0.0497	0.0411	0.0316	0.0234	0.0167
120	0.0998	0.0632	0.0495	0.0419	0.0371	0.0337	0.0312	0.0292	0.0279	0.0266	0.0218	0.0165	0.0120	0.0083
∞	0	0	0	0	0	0	0	0	0	0	0	0	0	0

(continued)

Table 4-4. (continued).

$\alpha = .01$

k \ ν	1	2	3	4	5	6	7	8	9	10	16	36	144	∞
2	0.9999	0.9950	0.9794	0.9586	0.9373	0.9172	0.8988	0.8823	0.8674	0.8539	0.7949	0.7067	0.6062	0.5000
3	0.9933	0.9423	0.8831	0.8335	0.7933	0.7606	0.7335	0.7107	0.6912	0.6743	0.6059	0.5153	0.4230	0.3333
4	0.9676	0.8843	0.7814	0.7212	0.6761	0.6410	0.6129	0.5897	0.5702	0.5536	0.4884	0.4057	0.3251	0.2500
5	0.9279	0.7885	0.6957	0.6329	0.5875	0.5531	0.5259	0.5037	0.4854	0.4697	0.4094	0.3351	0.2644	0.2000
6	0.8828	0.7218	0.6258	0.5635	0.5195	0.4866	0.4608	0.4401	0.4229	0.4084	0.3529	0.2858	0.2229	0.1667
7	0.8376	0.6644	0.5685	0.5080	0.4659	0.4347	0.4105	0.3911	0.3751	0.3616	0.3105	0.2494	0.1929	0.1429
8	0.7945	0.6152	0.5209	0.4627	0.4226	0.3932	0.3704	0.3522	0.3373	0.3248	0.2779	0.2214	0.1700	0.1250
9	0.7544	0.5727	0.4810	0.4251	0.3870	0.3592	0.3378	0.3207	0.3067	0.2950	0.2514	0.1992	0.1521	0.1111
10	0.7175	0.5358	0.4469	0.3934	0.3572	0.3308	0.3106	0.2945	0.2813	0.2704	0.2297	0.1811	0.1376	0.1000
12	0.6528	0.4751	0.3919	0.3428	0.3099	0.2861	0.2680	0.2535	0.2419	0.2320	0.1961	0.1535	0.1157	0.0833
15	0.5747	0.4069	0.3317	0.2882	0.2593	0.2386	0.2228	0.2014	0.2002	0.1918	0.1612	0.1251	0.0934	0.0667
20	0.4799	0.3297	0.2654	0.2288	0.2048	0.1877	0.1748	0.1646	0.1567	0.1501	0.1248	0.0960	0.0709	0.0500
24	0.4247	0.2871	0.2295	0.1970	0.1759	0.1608	0.1495	0.1406	0.1338	0.1283	0.1060	0.0810	0.0595	0.0417
30	0.3632	0.2412	0.1913	0.1635	0.1454	0.1327	0.1232	0.1157	0.1100	0.1054	0.0867	0.0658	0.0480	0.0333
40	0.2940	0.1915	0.1508	0.1281	0.1135	0.1033	0.0957	0.0898	0.0853	0.0816	0.0668	0.0503	0.0363	0.0250
60	0.2151	0.1371	0.1069	0.0902	0.0796	0.0722	0.0668	0.0625	0.0594	0.0567	0.0461	0.0344	0.0245	0.0167
120	0.1225	0.0759	0.0585	0.0489	0.0429	0.0387	0.0357	0.0334	0.0316	0.0302	0.0242	0.0178	0.0125	0.0083
∞	0	0	0	0	0	0	0	0	0	0	0	0	0	0

the possibility of an outlier exists. Another level of analysis may also be needed; for example, it may be necessary to apply some type of normalizing transformation (e.g., log Y_i). Cochran's test can be considered as a general test for equality of variance among several populations. If the homogeneity hypothesis is not rejected, then the variances can be pooled to provide a more reliable estimate of variances (that is, we obtain more degrees of freedom).

ANOVA is performed using the assumption of constant variance. A major pitfall which can invalidate the statistical analysis is a situation where one variance is large compared to the rest. Cochran's test, outlined below, is specifically designed to detect this type of deviation from homogeneous variance. The steps to follow in this analysis are as follows:

- Compute and tabulate the variances for each set of replicates. Note that this method requires that each variance must have the same number of observations.
- Choose the risk of assuming inhomogeneity when there is none (α).
- Table 4-4 provides critical values of Cochran's test. The values are given as a function of α, k-cells and ν degrees of freedom.
- Compute the critical value C^*:

$$C^* = (s_{i_{max}}^2)/ \sum_{i=1}^{k} s_i^2 \qquad (7)$$

- Apply the following criteria to draw conclusions from the analysis:
  - If $C^* > C$, then we are $100(1 - \alpha)\%$ confident that the variance is inhomogeneous.
  - If $C^* < C$, then there is insufficient evidence to conclude that the variance is inhomogeneous.

The following illustration demonstrates the analysis.

Illustration 7

The molecular weights of four rubber samples obtained during a period where the catalyst flowmeter's accuracy was questionable were compared to a control sample. The data are:

Rubber Samples

1	2	3	4	5 (control)
3174	3168	3175	3179	
3176	3171	3177	3174	3181
3175	3772	3177	3173	3179

- To ease computations we can code data by subtracting 3100 from each data value.
- $s_1^2 = 1.0$, $s_2^2 = 4.3$, $s_3^2 = 1.3$, $s_4^2 = 1.0$, $s_5 = 1.3$
- $C^* = 4.3/8.9 = 0.48$

From Table 4-4, $C_{0.99} = 0.7885$ at $\nu = 2$ and $k = 5$.

Since $C^* < C$, we cannot reject homogeneity.

As noted earlier, if homogeneity is not rejected, the variances can be pooled to obtain degrees of freedom. Another way of stating this is that when the standard deviation is independent of the factor level, the variance estimates at the varous levels can be pooled to obtain the best estimate of σ. This is done by first computing the sample variances for each of the k-levels of Y. The pooled standard deviation can be calculated from the following formula:

$$s_p = \sqrt{\sum_{i=1}^{k} n_i \, s_i^2 \Big/ \sum_{i=1}^{k} \nu_i} \qquad (8)$$

The degrees of freedom associated with s_p are:

$$\nu_p = \sum_{i=1}^{k} \nu_i \qquad (9)$$

One-Way Analysis of Variance (ANOVA)

ANOVA is similar to the hypothesis test on two means illustrated in Chapter 2. The form of the analysis is, however, slightly different. This statistical method is used to analyze a completely randomized single factor design. Key assumptions made are that the experimental error is *normally* distributed and it is identical across factor levels. The stepwise procedure for ANOVA is as follows:

- Select the risk of asserting that the factor effect exists when it actually does not (i.e., the α-risk).
- Tabulate the response.data, Y_{ij}, for $i = 1, \ldots , k$ factor levels. Include in this tabulation the number of replicates for each factor level, n_i.

For each factor level compute the following:

$$\text{Response Total, } T_i = \sum_{j=1}^{n} Y_{ij} \qquad (10)$$

$$\text{Response Average, } \bar{Y}_i = T_i/n_i \qquad (11)$$

$$\text{Sum of Squared Responses, } \sum_{j=1}^{n_i} Y_{ij}^2 \tag{12}$$

- Compute the following by summing across the k factor levels:

$$\text{Grand Total, } T = \sum_{i=1}^{k} T_i \tag{13}$$

$$\text{Total Experiments, } N = \sum_{i=1}^{k} n_i \tag{14}$$

$$\text{Total Sum of Squared Responses, } \sum_{i=1}^{k} \sum_{j=1}^{n_i} Y_{ij}^2 = \sum_{ij} Y_{ij}^2 \tag{15}$$

- Compute the quantities:

$$\sum_{i=1}^{k} \left(\frac{T_i^2}{n_i} \right) \text{ and } \frac{T^2}{N} \text{ (Global Average)} \tag{16}$$

- Compute the *error sum of squares:*

$$SSE = \sum_{ij} Y_{ij}^2 - \sum_{i} \left(\frac{T_i^2}{n_i} \right) = \sum_{ij} (Y_{ij} - \bar{Y}_i)^2 \tag{17}$$

This quantity is essentially a measure of the experimental error.

- Compute the *factor sum of squares:*

$$SSA = \sum_{i} \left(\frac{T_i^2}{n_i} \right) - \frac{T^2}{N} = \sum_{i} n_i (\bar{Y}_i - \bar{\bar{Y}})^2 \tag{18}$$

This quantity reflects the experimental error plus the factor effect.

- Compute the *mean squares*, which is simply the sum of squares divided by the degrees of freedom, i.e.,

$$MSE = SSE/\nu_E, \text{ where } \nu_E = N - k$$

$$MSA = SSA/\nu_A, \text{ where } \nu_A = k - 1 \tag{19}$$

- Compute the F^* statistic as follows:

$$F^* = \text{MSA}/MSE \qquad (20)$$

- From Table 4-2 obtain the F-value for $100(1 - \alpha)\%$ confidence and ν_A, ν_E degrees of freedom.
- The following criteria are applied to draw a conclusion from the analysis:
  - If $F^* > F$, we may assert that the factor effect exists with $100(1 - \alpha)\%$ confidence.
  - If $F^* < F$, we conclude that there is insufficient evidence to support the assertion that the factor effect exists.

The following problem is taken from Miller (1985). The illustration demonstrates application of the statistical method to a quality control problem.

Illustration 8

In the manufacture of plastic pipe, the melted plastic is extruded through an annular space between a die body and a mandrel. As the pipe is withdrawn from the die it is air cooled and cut into specified lengths. In addition to the length, the pipe must meet manufacturing specifications for wall thickness (minimum and maximum). The perfect part would naturally have uniform thickness and hence be perfectly concentric. Extruder variables that can affect manufacturing are: eccentricity of the die and mandrel, and machine surging. It is important therefore, to have a quantitative understanding of how the operating variability of the equipment impacts on the quality of the extruded part.

The study by Miller (1985) evaluates the degree of control in such a plant by obtaining wall thickness measurements of extruder pipe made from two similar machines [a newly conditioned extruder (Extruder A) and an older unit of the same design (Extruder B)]. Ten samples of pipe from each extruder were obtained, and wall thickness was measured at 8 evenly spaced locations moving clockwise around the circumference.

The data reported by Miller (1985) was tabulated on a Lotus 1-2-3 spreadsheet and is shown on the left-hand side of Table 4-5.

The right-hand side (rhs) of Table 4-5 shows the calculations performed in the analysis. The sums of squared deviation (ss) are calculated as:

$$ss_{Total} = \sigma^2(n - 1) = \Sigma(X_{ij}^2) - n\bar{\bar{X}}^2 \qquad (21)$$

where $\bar{\bar{X}}$ = global average. It may also be computed by:

$$ss_{total} = \Sigma(x_{ij}^2) - (\text{Global Total})^2/n \qquad (22)$$

Table 4-5. Summary of ANOVA Calculations for Illustration 8.

Wall Thickness Measurements for Extruder A

Sample	1	2	3	4	5	6	7	8	Total	Average	SS
1	0.142	0.145	0.143	0.14	0.137	0.137	0.139	0.137	0.119	0.14	0.156866
2	0.143	0.146	0.141	0.143	0.136	0.138	0.139	0.141	0.126	0.140875	0.158837
3	0.141	0.145	0.142	0.142	0.135	0.136	0.134	0.139	0.113	0.13925	0.155232
4	0.14	0.143	0.142	0.138	0.137	0.133	0.138	0.139	0.109	0.13875	0.15408
5	0.142	0.145	0.143	0.142	0.136	0.136	0.141	0.141	0.125	0.14075	0.158556
6	0.141	0.144	0.142	0.142	0.137	0.136	0.14	0.14	0.121	0.14025	0.15741
7	0.137	0.142	0.141	0.138	0.138	0.14	0.134	0.14	0.109	0.13875	0.154058
8	0.14	0.142	0.14	0.142	0.135	0.14	0.139	0.139	0.116	0.139625	0.155995
9	0.143	0.146	0.145	0.14	0.135	0.137	0.139	0.138	0.122	0.140375	0.157749
10	0.143	0.147	0.145	0.142	0.142	0.137	0.138	0.141	0.134	0.141875	0.161105
Total	0.125	0.158	0.137	0.122	0.081	0.083	0.094	0.081	−0.093		
Average	0.1412	0.1445	0.1424	0.1409	0.1368	0.137	0.1381	0.1395		0.14005	

$X_j^2 =$

Total	1.569888
SS-Total	0.000767
SS-Columns	0.000513
SS-Residual	0.000254
Mean Square Residual	0.000073
Residual	0.000003
F	145.3018

(continued)

Table 4-5. (continued).

Wall Thickness Measurements for Extruder B

Sample	1	2	3	4	5	6	7	8	Total	Average	SS
1	0.146	0.144	0.14	0.134	0.138	0.141	0.145	0.145	1.133	0.141625	0.160583
2	0.161	0.16	0.155	0.151	0.153	0.158	0.163	0.163	1.264	0.158	0.199858
3	0.153	0.159	0.152	0.151	0.153	0.159	0.163	0.156	1.246	0.15575	0.19419
4	0.144	0.146	0.138	0.137	0.139	0.143	0.147	0.141	1.135	0.141875	0.161125
5	0.162	0.159	0.153	0.149	0.152	0.156	0.161	0.161	1.253	0.156625	0.196417
6	0.152	0.154	0.145	0.145	0.148	0.153	0.153	0.149	1.199	0.149875	0.179793
7	0.139	0.139	0.132	0.132	0.134	0.139	0.141	0.137	1.093	0.136625	0.149417
8	0.163	0.16	0.162	0.156	0.152	0.149	0.15	0.158	1.25	0.15625	0.195518
9	0.159	0.158	0.149	0.149	0.151	0.157	0.16	0.157	1.24	0.155	0.192346
10	0.149	0.149	0.143	0.141	0.145	0.149	0.151	0.146	1.173	0.146625	0.172075
Total	0.223	0.223	0.164	0.14	0.16	0.199	0.229	0.208	1.546		1.801322
Average	0.1528	0.1528	0.1469	0.1445	0.1465	0.1504	0.1513		0.149825		

$X^2_i =$

Total 1.801322

SS-Total 0.005519

SS-Columns 0.000809

SS-Residual 0.00471

Mean Square 0.000115

Residual 0.000065

F 12.37528

The row denoted ss_{Total} on the rhs of Table 4-5 is simply the sum of the rows minus the global mean, i.e.,

$$ss_{Total} = (-1)^2 + (2)^2 + \ldots + (-2)^2 - 80(-2.95)^2 = 767.8$$

Similarly, the ss attributable to differences in position (i.e., to the different points along the pipe's circumference) is computed from the averages of the eight points:

$$ss_{columns} = r\Sigma(\bar{X}_j^2) - n\bar{\bar{X}}^2 \qquad (23)$$

where $\bar{X}_j$ = average for column j
r = number of rows (i.e., of points in a column).

That is, the column ss is the squared deviation attributed to the differences between the columns. Physically, the variation is due to measurements taken at different positions along the circumference. So, the calculation on the lhs of Table 4-5 is:

$$ss_{columns} = 10((-1.8)^2 + (1.5)^2 + \ldots + (-3.5)^2) - 80(-2.95)^2 = 513.4$$

The deviation within columns is defined as the residual:

$$ss_{residual} = ss_{total} - ss_{column} \qquad (24)$$

The mean squared deviations are the variances (i.e., column and residual). The ratio between the column mean squared deviations and that of the residual constitutes an F-ratio. That is, the F-value is the ratio between the explained and unexplained variances. For a large F, the factor is significant (the level of significance is the probability value, P).

From the calculations in Table 4-5, Extruder A shows that the location of measurement along the circumference is important. We might therefore consider some die adjustments to make Extruder A's production more uniform.

A second illustration should increase the reader's expertise in ANOVA.

Illustration 9

We wish to construct a capacitance probe to be used in an experimental fluid bed to study bubble phenomena (the change in bed capacitance reflects the presence of bubbles). Several probe geometries are being evaluated by immersing the probe in a static bed of fine particles. The objective of this analysis is to evaluate whether the sensor geometry influences the measurement.

The capacitance measurements (picoFarads) obtained from four geometries are:

Probe Configuration

	I	II	III	IV	
Capacitance (pF)	30	34	25	32	
	36	41	26	29	
	42	47	25	35	
	29	50	29	43	
	40	51	33	41	

Total	177	223	138	180	718 = Grand Total
n_i	5	5	5	5	20 = N
$\bar{Y}_i$		35.4	44.6	27.6	36.0	35.9 = $\bar{Y}$ (Global Average)
$\displaystyle\sum_j Y_{ij}^2$		6401	10147	3856	6620	27024 = $\displaystyle\sum_{ij} Y_{ij}^2$

$$\sum_i \frac{T_i^2}{n_i} = \frac{(177)^2}{5} + \frac{(223)^2}{5} + \frac{(138)^2}{5} + \frac{(180)^2}{5} = 26500$$

$$\frac{\text{Grand Total}^2}{N} = \frac{(718)^2}{20} = 25776$$

Computed error sum of squares:

$$\text{SSE} = \sum_{ij} Y_{ij}^2 - \sum_i \frac{T_i^2}{n_i} = 27024 - 26500 = 523.6$$

Computed factor sum of squares:

$$\text{SSA} = \sum_i \frac{T_i^2}{n_i} - \frac{T^2}{N} = 26500 - 25776 = 724$$

Computed Means Squares:

$$\nu_E = N - k = 20 - 4 = 16$$

$$\text{MSE} = \frac{\text{SSE}}{\nu_E} = \frac{523.6}{16} = 32.7$$

$$\nu_A = k - 1 = 4 - 1 = 3$$

$$MSA = \frac{SSA}{\nu_A} = \frac{724}{3} = 241.3$$

Compute F^*:

$$F^* = \frac{MSA}{MSE} = \frac{241.3}{32.7} = 7.38$$

From Table 4-2 for $\alpha = 0.05$, $F_{0.95}(3,16) = 3.24$.

$F^* > F$, therefore, we may conclude that the probe geometry influences the sensitivity of the measurement.

We may summarize the important computations for this illustration of ANOVA as follows:

Source of Variations	Degrees of Freedom	SS	MS	F^*
Probe Geometry	3	724	241.3	7.38
Error	16	524	32.7	

$F_{0.95}(3,16) = 3.24$.

Two- and Three-Way ANOVA's

ANOVA can be readily applied to separate out the effects of two or more variables. In principle, the number of variables that can be studied in a system is infinite. However, in practice, when the number of variables exceeds three, the computations become too cumbersome. The application of two-way ANOVA is demonstrated in the following illustration.

Illustration 10

Continuing with the problem presented by Miller (1985) in Illustration 8, we wish to separate the effects of the die-and-mandrel geometry and of machine surging. In terms of the tabulations presented in Table 4-5, what we need is an additional ss calculation, except this time for the rows of data:

$$SS_{Rows} = c\Sigma(\bar{X}_i^2) - n\bar{\bar{X}}^2 \tag{25}$$

where $\bar{X}_i$ = row averages
      c = number of columns (i.e., number of data points in a row).

Equation (25) may also be expressed in terms of the row and global totals:

$$SS_{Rows} = (\Sigma(\text{Row Total})^2/c) - (\text{Global Total})^2/n \tag{26}$$

This calculation represents time in the data.

Hence, the calculations for Extruder A are:

$$SS_{Rows} = 8\{(0.14)^2 + (0.1409)^2 \ldots + (0.1419)^2\} - 80(.14005)^2 = 7.08 \times 10^{-5}$$

$$SS_{Residual} = SS_{Total} - SS_{Columns} - SS_{Rows} = 7.67 \times 10^{-4} - 5.13 \times 10^{-4}$$

$$- 7.08 \times 10^{-5} = 1.832 \times 10^{-4}$$

Similar calculations are carried out for Extruder B and we can summarize the two-way ANOVA for this problem (using tabulations from Table 4-5) as follows:

Extruder A

Source	SS	Degree of Freedom	MS	F	P
Total	7.67×10^{-4}	79	−	−	−
Position (columns)	5.13×10^{-4}	7	7.3×10^{-5}	25.1	$< <0.001$
Time (Rows)	7.08×10^{-5}	9	7.87×10^{-6}	2.70	0.01
Residual	1.832×10^{-4}	63	2.91×10^{-6}	−	−

Extruder B

Source	SS	Degree of Freedom	MS	F	P
Total	5.519×10^{-3}	79	−	−	−
Position (columns)	8.09×10^{-4}	7	1.156×10^{-4}	15.81	$< <0.001$
Time (Rows)	4.2493×10^{-3}	9	4.721×10^{-4}	64.6	$< <0.001$
Residual	4.607×10^{-4}	63	7.31×10^{-6}	−	−

The above values indicate that both position and time (i.e., surging) significantly contribute to the variance for both machines. For Extruder A, position is a greater source of unsteady operation, whereas for Extruder B, surging appears to be a greater source than position (although positioning or lack of centering the die-mandrel arrangement is certainly significant).

Illustration 11

This example illustrates application of a three-way ANOVA for assessing the effects of process variables on gel formation in a polymer made via a solution polymerization process. Polymer samples were made under the following conditions:

Experiment 1 – Base case polymer made (reactor lined out).
Experiment 2 – The base modifier (controlling the pH of the solution) was lowered.
Experiment 3 – Conditions returned to base case polymer and alkyl rate lowered.
Experiment 4 – Normal production requires quenching the reactor product to stop the reaction. In this experiment no quenching was done.

The polymers were then extruded into a thin film and the number of gel particles per square inch of film were counted. The results are reported in the form of coded data below, as 1–Relative Gel Count (where the relative count is the ratio of measured particles to the mean for the control polymer).

Run	Low Base	Low Alkyl	No Quench	Total
1	0.03	0.60	0.58	1.21
	-0.95	0.45	0.68	0.18
2	-1.05	0.48	0.75	0.18
	-1.61	0.56	0.76	-0.29
Total	-3.58	2.09	2.77	1.28

The extrusion tests were performed twice, and two different lengths of film were analyzed for each run. The ANOVA table is prepared in the usual manner, i.e., sums of squares (SS) and mean squares (MS) are calculated, but here we have duplicates, and so we need an additional variance calculation. This is the SS for differences between pairs, which is termed SS_{Error}. In general, it is computed by averaging the replicates and summing the squared deviations from this average of the individual measurements. For this case of duplicates, a simplified expression is:

$$SS_{Error} = (1/2)((x_{11} - x_{12})^2 + (x_{21} - x_{22})^2 \ldots + (x_{n1} - x_{n2})^2)$$

Hence, $SS_{Error} = 0.30$.

And by calculating the sums of squares:

$$SS_{Total} = (0.03)^2 + (-0.95)^2 \ldots + (0.76)^2 - (1.28)^2/12 = 7.50.$$

$$SS_{Base} = (0.03)^2 + (-0.95)^2 + (-1.05)^2 + (-1.61)^2 - (1.28)^2/12 = 4.46.$$

$$SS_{Alkyl} = 0.97$$

$SS_{Quench} = 1.80$

$SS_{Residual} = 7.50 - 4.46 - 0.97 - 1.80 = 0.27.$

The ANOVA summary is:

Source	SS	Degrees of Freedom	MS	F	F⁺
Total	7.50	11	—	—	—
Base	4.46	3	1.49	10.64	29.8
Alkyl	0.97	3	0.32	2.29	6.4
Quench	1.80	3	0.60	4.29	12.0
Error⁺	0.30	6	0.05	—	—
Residual	0.27	2	0.14	—	—

Note that we have computed an additional F-value based on the Error MS. The degrees of freedom for the Error MS are simply the number of pairs. Both F-values arrive at the same conclusion, although the calculation based on error makes a stronger statement—namely, that the base modifier has a dominant role in gel formation. We may also conclude that quenching is important, and that the alkyl concentration has a lesser effect.

We may take this analysis a step further by asking the question: how does the response vary with the significant factor level? The following steps apply to this analysis:

- Tabulate the k means ($\bar{Y}_i$) in ascending order.
- Enter the ANOVA table to take the MSE and ν_E.
- Compute the standard error of the mean:

$$S\bar{y} = \sqrt{MSE/n}$$

  where $n = N/k$.
- Use Table 4-6 (Studentized ranges for different α-values). For the appropriate α-value use the row entry for the error degrees of freedom and obtain values of the studentized ranges for $p = 2,3, \ldots k$.
- Multiply the ranges by $S\bar{y}$ to obtain the least significant ranges (LSR).
- Compare the differences among means with the LSR's in the following manner:
  −Compare the extreme means with LSR for $p = k - 1$.
  −Compare means $k - 1$ apart with LSR for $k - 1$.
  −Compare up to the adjacent means using LSR for $p = 2$.

The following example illustrates the method.

Table 4-6. Studentized Ranges.

$\alpha = 0.10$

ν_E \ P	2	3	4	5	6	7	8	9	10
1	8.93	13.44	16.36	18.49	20.15	21.51	22.64	23.62	24.48
2	4.13	5.73	6.77	7.54	8.14	8.63	9.05	9.41	9.72
3	3.33	4.47	5.20	5.74	6.16	6.51	6.81	7.06	7.29
4	3.01	3.98	4.59	5.03	5.39	5.68	5.93	6.14	6.33
5	2.85	3.72	4.26	4.66	4.98	5.24	5.46	6.56	5.82
6	2.75	3.56	4.07	4.44	4.73	4.97	5.17	5.34	5.50
7	2.68	3.45	3.93	4.28	4.55	4.78	4.97	5.14	5.28
8	2.63	3.37	3.83	4.17	4.43	4.65	4.83	4.99	5.13
9	2.59	3.32	3.76	4.08	4.34	4.54	4.72	4.87	5.01
10	2.56	3.27	3.70	4.02	4.26	4.47	4.64	4.78	4.91
11	2.54	3.23	3.66	3.96	4.20	4.40	4.57	4.71	4.84
12	2.52	3.20	3.62	3.92	4.16	4.35	4.51	4.65	4.78
13	2.50	3.18	3.59	3.88	4.12	4.30	4.46	4.60	4.72
14	2.49	3.16	3.56	3.85	4.08	4.27	4.42	4.56	4.68
15	2.48	3.14	3.54	3.83	4.05	4.23	4.39	4.52	4.64
16	2.47	3.12	3.52	3.80	4.03	4.21	4.36	4.49	4.61
17	2.46	3.11	3.50	3.78	4.00	4.18	4.33	4.46	4.58
18	2.45	3.10	3.49	3.77	3.98	4.16	4.31	4.44	4.55
19	2.45	3.09	3.47	3.75	3.97	4.14	4.29	4.42	4.53
20	2.44	3.08	3.46	3.74	3.95	4.12	4.27	4.40	4.51
24	2.42	3.05	3.42	3.69	3.90	4.07	4.21	4.34	4.44
30	2.40	3.02	3.39	3.65	3.85	4.02	4.16	4.28	4.38
40	2.38	2.99	3.35	3.60	3.80	3.96	4.10	4.21	4.32
60	2.36	2.96	3.31	3.56	3.75	3.91	4.04	4.16	4.25
120	2.34	2.93	3.28	3.52	3.71	3.86	3.99	4.10	4.19
∞	2.33	2.90	3.24	3.48	3.66	3.81	3.93	4.04	4.13

(continued)

Table 4-6. (continued).

$\alpha = 0.05$

ν_E \ P	2	3	4	5	6	7	8	9	10
1	17.97	26.98	32.82	37.08	40.14	43.12	45.40	47.36	49.07
2	6.08	8.33	9.80	10.88	11.74	12.44	13.03	13.54	13.99
3	4.50	5.91	6.82	7.50	8.04	8.48	8.85	9.18	9.46
4	3.93	5.04	5.76	6.29	6.71	7.05	7.35	7.60	7.83
5	3.64	4.60	5.22	5.67	6.03	6.33	6.58	6.80	6.99
6	3.46	4.34	4.90	5.30	5.63	5.90	6.12	6.32	6.49
7	3.34	4.16	4.68	5.06	5.36	5.61	5.82	6.00	6.16
8	3.26	4.04	4.53	4.89	5.17	5.40	5.60	5.77	5.92
9	3.20	3.95	4.41	4.76	5.02	5.24	5.43	5.59	5.74
10	3.15	3.88	4.33	4.65	4.91	5.12	5.30	5.46	5.60
11	3.11	3.82	4.26	4.57	4.82	5.03	5.20	5.35	5.49
12	3.08	3.77	4.20	4.51	4.75	4.95	5.12	5.27	5.39
13	3.06	3.73	4.15	4.45	4.69	4.88	5.05	5.19	5.32
14	3.03	3.70	4.11	4.41	4.64	4.83	4.99	5.13	5.25
15	3.01	3.67	4.08	4.37	4.59	4.78	4.94	5.08	5.20
16	3.00	3.65	4.05	4.33	4.56	4.74	4.90	5.03	5.15
17	2.98	3.63	4.02	4.30	4.52	4.70	4.86	4.99	5.11
18	2.97	3.61	4.00	4.28	4.49	4.67	4.82	4.96	5.07
19	2.96	3.59	3.98	4.25	4.47	4.65	4.79	4.92	5.04
20	2.95	3.58	3.96	4.23	4.45	4.62	4.77	4.90	5.01
24	2.92	3.53	3.90	4.17	4.37	4.54	4.68	4.81	4.92
30	2.89	3.49	3.85	4.10	4.30	4.46	4.60	4.72	4.82
40	2.86	3.44	3.79	4.04	4.23	4.39	4.52	4.63	4.73
60	2.83	3.40	3.74	3.98	4.16	4.31	4.44	4.55	4.65
120	2.80	3.36	3.68	3.92	4.10	4.24	4.36	4.47	4.56
∞	2.77	3.31	3.63	3.86	4.03	4.17	4.29	4.39	4.47

(continued)

Table 4-6. (continued).

$\alpha = 0.01$

ν_E \ P	2	3	4	5	6	7	8	9	10
1	90.03	135.0	164.3	185.6	202.2	215.8	227.2	237.0	245.6
2	14.04	19.02	22.29	24.72	26.63	28.20	29.53	30.68	31.69
3	8.26	10.62	12.17	13.33	14.24	15.00	15.64	16.20	16.69
4	6.51	8.12	9.17	9.96	10.58	11.10	11.55	11.93	12.27
5	5.70	6.98	7.80	8.42	8.91	9.32	9.67	9.97	10.24
6	5.24	6.33	7.03	7.56	7.97	8.32	8.61	8.87	9.10
7	4.95	5.92	6.54	7.01	7.37	7.68	7.94	8.17	8.37
8	4.75	5.64	6.20	6.62	6.96	7.24	7.47	7.68	7.86
9	4.60	5.43	5.96	6.35	6.66	6.91	7.13	7.33	7.49
10	4.48	5.27	5.77	6.14	6.43	6.67	6.87	7.05	7.21
11	4.39	5.15	5.62	5.97	6.25	6.48	6.67	6.84	6.99
12	4.32	5.05	5.50	5.84	6.10	6.32	6.51	6.67	6.81
13	4.26	4.96	5.40	5.73	5.98	6.19	6.37	6.53	6.67
14	4.21	4.89	5.32	5.63	5.88	6.08	6.26	6.41	6.54
15	4.17	4.84	5.25	5.56	5.80	5.99	6.16	6.31	6.44
16	4.13	4.79	5.19	5.49	5.72	5.92	6.08	6.22	6.35
17	4.10	4.74	5.14	5.43	5.66	5.85	6.01	6.15	6.27
18	4.07	4.70	5.09	5.38	5.60	5.79	5.94	6.08	6.20
19	4.05	4.67	5.05	5.33	5.55	5.73	5.89	6.02	6.14
20	4.02	4.64	5.02	5.29	5.51	5.69	5.84	5.97	6.09
24	3.96	4.55	4.91	5.17	5.37	5.54	5.69	5.81	5.92
30	3.89	4.45	4.80	5.05	5.24	5.40	5.54	5.65	5.76
40	3.82	4.37	4.70	4.93	5.11	5.26	5.39	5.50	5.60
60	3.76	4.28	4.59	4.82	4.99	5.13	5.25	5.36	5.45
120	3.70	4.20	4.50	4.71	4.87	5.01	5.12	5.21	5.30
∞	3.64	4.12	4.40	4.60	4.76	4.88	4.99	5.08	5.16

Illustration 12

Continuing with the previous problem, we decide to explore the effect of the base modifier at three concentrations:

	Low	Medium	High		
	0.03	0.38	-0.79		
	-0.95	0.24	-0.58		
	-1.05	0.69	-0.80		
	-1.61	-0.16	-0.48		
	-0.51	0.10	-0.77		
T_i	-4.09	1.25	-3.42	-6.26	Grand Total
n_i	5	5	5	15	N
$\bar{Y}_i$	-0.82	0.25	-0.68	-0.42	$\bar{Y}$
$\sum_i Y_{ij}^2$	4.86	0.71	2.42	7.99	$\sum_{ij} Y_{ij}^2$

$$\sum_i \frac{T_i^2}{n_i} = \frac{(-4.09)^2}{5} + \frac{(1.25)^2}{5} + \frac{(-3.42)^2}{5} = 6.00$$

$$\frac{(\text{Grand Total})^2}{N} = \frac{(-6.26)^2}{15} = 2.61$$

$$SSE = \sum_{ij} Y_{ij}^2 - \sum_i T_i^2/n_i = 7.99 - 6.00 = 0.99$$

$$SSA = \sum_i \frac{T_i^2}{n_i} - \frac{(\text{Grand Total})^2}{N} = 6 - 2.61 = 3.39$$

$$\nu_E = N - k = 15 - 3 = 12$$

$$MSE = \frac{SSE}{\nu_E} = \frac{0.99}{12} \doteq 0.08$$

$\bar{Y}_3$	$\bar{Y}_2$	$\bar{Y}_1$
-0.82	-0.68	0.25

$$S_{\bar{y}} = \sqrt{\frac{MSE}{n}} = \sqrt{\frac{0.08}{5}} = 0.13$$

From Table 4-6, $\alpha = 0.01$, $\nu_E = 12$:

p	2	3
ranges	4.32	5.05
LSR	0.56	0.66

For $p = 3$

$$\text{LSR}$$

$$0.25 - (-0.82) = 1.07 \quad \overline{0.66} \quad \text{Diff.}$$

For $p = 2$

$$0.25 - (-0.68) = 0.93 \quad 0.56 \quad \text{Diff.}$$

We may conclude that all three levels are significantly different.

Illustration 13

This last problem demonstrates a three-way ANOVA with an interaction effect between parameters. The example is concerned with evaluating reactor technologies for the production of methane gas from coal. Four different reactor configurations were studied: a fluidized single-screw mixer retort, a fluidized dual-screw mixer retort, a conventional fluidized bed, and a centrifugal fluid bed. The first variable studied was pressure; that is, each reactor process was operated at two pressure levels but at constant reactor temperature. The other variable was the point of addition of recycle syngas. On each unit the recycle gas was added to the lead reactor/retort (RI) then to a surge tank (R2) which followed each retort. Thus, the surge tank served as a second reactor in series. The experimental design was therefore comprised of 16 combinations (i.e., $4 \times 2 \times 2$) of reactor configuration, pressure and point of recycle gas injection. The overall methane yields in ppt from total syngas production from each system are as follows:

Reactor Configuration	Recycle to R1		Recycle to R2		Total
	250 psi	700 psi	250 psi	700 psi	
1	410	611	422	499	1,962
2	215	370	479	701	1,765
3	295	301	409	493	1,498
4	615	600	515	730	2,960
Total	1,535	1,882	1,845	2,423	7,685

Calculations of the sums of squares when two variables interact are:

$SS_{total} = (410)^2 + (215)^2 \ldots + (730)^2 - (7685)^2/16 = 322{,}857$

$SS_{conf.} = ((1962)^2 \ldots + (2460)^2)/4 - (7685)^2/16 = 123{,}866$

$SS_{recycle} = ((1{,}535 + 1{,}882)^2 + (1{,}845 + 2{,}423)^2)/8 - (7{,}685)^2/16 = 45{,}263$

$SS_{press.} = ((1535 + 1845)^2 + (1882 + 2423)^2)/8 - (7685)^2/16 = 53{,}477$

$SS_{conf./rec.} = ((410 + 611)^2 \ldots + (515 + 730)^2)/2 - (7685)^2/16 = 76{,}093$

$SS_{residual} = 322{,}857 - 123{,}866 - 45{,}263 - 53{,}477 - 76{,}093 = 24{,}158$

Note that the interaction SS is computed like the error SS for a test involving replicates. The three-way ANOVA computations with an interaction effect may be summarized as follows:

Source	SS	Degrees of Freedom	MS	F	P
Total	322,857	15	—	—	0.01
Configuration	123,866	3	41,289	11.96	<0.05
Recycle Loc.	45,263	1	45,263	13.12	<0.05
Pressure	53,477	1	53,477	15.50	<0.05
Conf. × Recy.	76,093	3	25,364	7.35	<0.5
Residual	24,158	7	3,451	—	

In this example, the ANOVA indicates that all the main effects (reactor configuration, pressure and point of recycle gas injection), are significant. The interaction between retort configuration and recycle gas injection also is significant. This implies that one mode of recycle is better for some configurations, while the other mode is better for others. Further exploration of all four arrangements should be made to fully establish the capabilities of each technology.

PRACTICE PROBLEMS

1   A less expensive catalyst has replaced our titanium-based catalyst in the production of a polymer used in blown sponge applications. Swell is an important property for the end-use application. On the following page are die swell measurements obtained from a capillary rheometer for the drop-in prototype. The control polymer is a

representative sample produced from the original catalyst system. Is the product variability the same for both polymers? Most customers extrude the product at shear rates > 1000 sec^{-1}. Are the majority of customers unlikely to detect a change in the polymer swell?

	SHEAR RATE (sec^{-1})			
	10	100	1000	5000
Drop-in Prototype Swell (%)	13	17	21	17
	12	22	17	10
	12	15	25	10
	10	19	15	15
	13	22	25	15
	14	23	17	10
	14	20	21	17
Control Polymer Swell (%)	6	13	15	18
	8	12	18	16
	11	11	14	19

2 We have two fluid bed driers operating in parallel to dry french fries. Fluid bed 2 (FB2) is a simulation of the plant and FB1 incorporates some proposed design modifications. The following moisture levels were measured from batches made in each dryer. Does FB1 improve the drying operation?

				Moisture (wt. %)					
FB1	2	3	1.5	5.2	6.3	4.7	3.2	3.7	4.1	4.2
FB2	4.7	5.3	0.5	6.2	5.3	4.6	4.3	3.5	2.0	6.9

3 Two reactor technologies are being considered for a commercial gasification process. The syngas yields from each pilot reactor are reported below. Does one of the reactors give an improved yield? How many measurements on each reactor system would be required to show up to a 5% yield increase or better?

Technology 1 Yields (%) 87, 85, 95, 93, 73, 72, 83, 91, 93, 92
Technology 2 Yields (%) 92, 90, 71, 69, 87, 98, 95, 93, 89, 91

4 Based on our production data, we shipped a customer 3 tons of polymer with a mean Mooney viscosity of 55 with $\sigma = 5.5$. Our production rate was 0.3 tons/hr and Mooney viscosity was measured every 2 hours during the run. The customer ran his own quality

control check as the crates of polymer were unloaded, resulting in the following Mooney viscosities (45, 49, 57, 53, 57.5, 49.5, 52, 56, 51, 50). Is there a possibility that a discrepancy exists in the measurements between us.

5 Heat of fusion data were obtained from 5 pads prepared from a control polymer annealed for 2 days. Seven measurements were then made on the same control, but this time the sample was annealed for 1 week. Are the variances for the two annealing times different if the variance estimate was $s_1^2 = 16.0$ and $s_2^2 = 9.8$? Assume 90% confidence in the analysis.

6 (A) The regular technician ran 12 assays on a catalyst make-up solution with a variance estimate of $s_1^2 = 1.0$. A temporary man ran 8 assays on the same sample with a variance estimate of $s_2^2 = 1.5$. Are the variances for the two men different?

(B) How many standards would each technician have to run in order to be able to pick up a 3:1 ratio of variances? For the statistical basis, choose $\alpha = 0.05$, single-sided comparison and $\beta = 0.50$.

7 Ten catalyst residue measurements (Vanadium levels, ppm) were made on each of two polymer batches giving average values of 15 and 12 ppm Va, respectively. Assuming that the standard deviation of individual Va determinations is 1%, can we say that the batches are different with 95% confidence?

8 Apply Cochran's test for equal variance to the following data.

$x_1 : Y_1 = 23, 19, 17, 24, 22, 19$
$x_2 : Y_2 = 6, 11, 15, 12, 7, 13$
$x_3 : Y_3 = 35, 37, 33, 32, 31, 29$
$x_4 : Y_4 = 29, 25, 28, 31, 31, 27$

9 Fertilizer pellets prepared in spray driers are loaded into drums via a weigh hopper. Some drums appear to have excessive fines, so we want to know whether the fertilizer was segregating during loading operations or whether the spray driers are producing smaller than expected cut sizes. To address this problem, single samples of fertilizer were retrieved from four different levels of five drums in sequence, and were screened to determine the weight percent of particles larger than 30 mesh. Use the data reported on the next page to establish within-drum and between-drum variations, and then subject the problem to ANOVA. Make recommendations as to where corrective actions should be taken in the process.

Weight % Larger Than 30 Mesh

| Drum | Level in Drum | | | | |
	1	2	3	4	Total
1	73.9	83.1	83.1	79.0	319.1
2	65.7	75.6	86.9	83.0	311.2
3	62.4	79.2	72.9	78.0	292.5
4	60.5	74.2	79.9	85.2	299.8
5	69.2	76.7	81.6	80.3	307.2
Total	331.7	388.2	404.4	405.5	1529.8

CHAPTER 5

The Correlation Coefficient and Introduction to Spectral Density Analysis

INTRODUCTION TO CORRELATION CONCEPTS

The terminology covered in this chapter is critical to the discussions and illustrations presented in the remaining sections of this volume. We now begin to encounter an important class of statistical problems, namely, correlation problems. A correlation problem addresses the joint variability of two populations with the aim of assessing whether they are related. The analysis is based on the assumption that the observations occur in pairs.

The first tool available for making an assessment of a possible correlation is a *scatter diagram*, which is a plot of one population versus another. Such a plot provides an indication as to the probable form and nature of the relationship. As examples, Figure 5-1a shows the cure properties of several polymers made in the plant versus equivalent grades made in a pilot scale reactor of 1/10 capacity. Based on these observations, the scatter plot indicates that the full-size and pilot units have been properly scaled to produce equivalent products. Figure 5-1b shows yet another example. In this case, the cure state (i.e., tightness of cure) of a polymer is shown plotted against the fraction of low molecular weight component found in the molecular weight distribution (expressed as % oligomers). The plot suggests a correlation between the amount of low molecular weight component and cure properties, i.e., the more low molecular weight components in the polymer, the poorer the cure.

When the data show a linear dependency between two characteristic parameters, we can quantify the relationship through the use of the *correlation coefficient*. That is, when the average value of one property (Y) appears to depend linearly on the value of a second property (X), then a correlation coefficient, r^* can be computed. The *population coefficient* is denoted by the symbol ϱ, and is estimated from r^*. The population correlation coefficient provides a quantitative measure of the degree of association between the two properties X and Y. The degree of association is represented graphically in Figure 5-2 for typical ϱ-values. The elliptical areas represent regions in which 95% of the population is found. To compute the correlation coefficient use the following procedure:

- Compute the sums of squares:

$$S_{xx} = \Sigma(x - \bar{x})^2 = \Sigma x^2 - (\Sigma x)^2/n \qquad (1A)$$

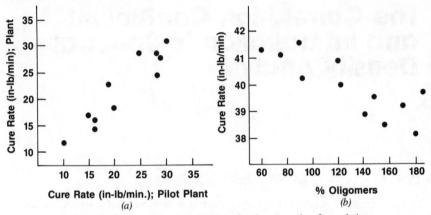

FIGURE 5-1. Examples of scatter plots showing strengths of correlations.

$$S_{yy} = \Sigma(y - \bar{y})^2 = \Sigma y^2 - (\Sigma y)^2/n \tag{1B}$$

$$S_{xy} = \Sigma(x - \bar{x})(y - \bar{y}) = \Sigma xy - (\Sigma x)(\Sigma y)/n \tag{1C}$$

- Compute the correlation coefficient:

$$r^* = S_{xy}/\sqrt{S_{xx}S_{yy}} \tag{2}$$

Review the following example.

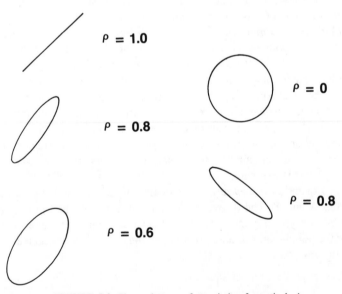

FIGURE 5-2. Shows degrees of association for typical ϱ's.

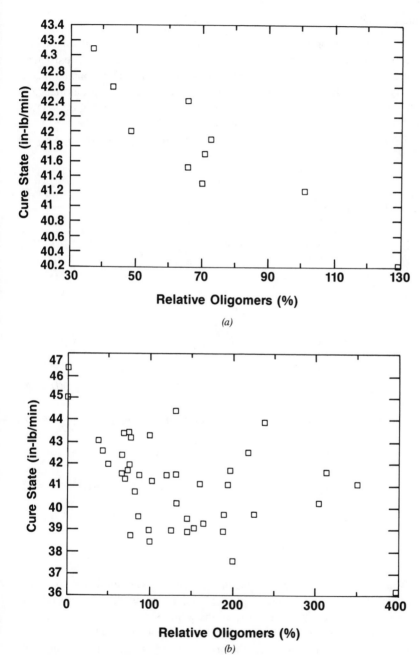

FIGURE 5-3. Scatter plots for Illustration 1.

Illustration 1

In establishing product property limits for our polymers we attempt to cor-relate cure performance with molecular weight distribution. In particular, there is some evidence suggesting that when the polymer contains a low molecular weight fraction (oligomers) a drop-off in cure performance is observed. This observation is suggested by the scatter plot in Figure 5-3a for polymers made with our standard catalyst.

Computing the correlation coefficient for the 10 data points shown in Figure 5-3a, we obtain:

i	x Oligomers (%)	y M^H (in-lb)	Σx	Σx^2	Σy	Σy^2	Σxy
1	38	43.1	38	1444	43.1	1857.6	1637.8
2	43	42.6	81	3293	85.7	3672.4	3469.6
3	48	42	•	•	•	•	•
4	65	42.5	•	•	•	•	•
5	65	41.5	•	•	•	•	•
6	69	41.3					
7	70	41.7					
8	71	41.7					
9	101	41.3					
10	129	40.2	699	5.559×10^4	417.9	1.747×10^4	2.903×10^4

$$S_{xx} = 5.554 \times 10^4 - (699)^2/10 = 6730.9$$

$$S_{yy} = 1.747 \times 10^4 - (417.9)^2/10 = 6.03$$

$$S_{xy} = 2.903 \times 10^4 - (699)(417.9)/10 = -181.2$$

$$r^* = -181.2/\sqrt{6730.9 \times 6.03} = -0.899$$

or $|r^*| = 0.899$

This is a high correlation coefficient suggesting a strong relation ($\varrho = 1$ is perfect). However, we would be justified in questioning whether r^* is reliable enough to provide a good estimate of the population correlation coefficient, ϱ. Figure 5-3b shows a scatter plot for the same polymer made with different catalysts. The r^* estimated from the original 10 measurements certainly does not represent an accurate estimate for the entire population. In fact, for the entire population, the computed r^* is:

$$S_{xx} = 1.1 \times 10^6 - (5.9 \times 10^3)/44 = 3.4 \times 10^5$$

$$S_{yy} = 7.5 \times 10^4 - (1.8 \times 10^3)/44 = 1.8 \times 10^2$$

$$S_{xy} = 2.4 \times 10^5 - (5.9 \times 10^3)(1.8 \times 10^3)/44 = 3.6 \times 10^3$$

$$\varrho \cong |r^*| = |-3.6 \times 10^3/\sqrt{(3.4 \times 10^5)(1.8 \times 10^2)}| = 0.46$$

We might very well wonder if a correlation actually exists with such a low correlation coefficient. In fact, the scatter plot of Figure 5-3b suggests that perhaps a non-linear relationship between M_H and oligomers exists.

The above illustrates that a large value of r^* may in fact be related to improper sampling of the population or perhaps entirely to chance selection. That is, it's possible to unknowingly select data sets which show a correlation while in fact, the population really has no correlation between X and Y (i.e., $\varrho = 0$). In the above example, the type catalyst may also be a second independent variable, such that both oligomers and catalyst type correlate with cure properties. The following is a quick procedure which tests for the case of $\varrho = 0$.

- Select α (i.e., the risk of claiming there is an association between X and Y when none exists.
- Use Table 5-1 to obtain r for a given α and n.
- Apply the following criteria:
  - If $|r^*| > r$, we can claim with $100(1 - \alpha)\%$ confidence that there is some association between X and Y. That is, we conclude that $\varrho \neq 0$.

Table 5-1. Percentiles of the Distribution of r when $\varrho = 0$.

n	$\alpha = .10$	$\alpha = .05$	$\alpha = .01$	n	$\alpha = .10$	$\alpha = .05$	$\alpha = .01$
5	.805	.878	.959	20	.378	.444	.561
6	.729	.811	.917	22	.360	.423	.537
7	.669	.754	.875	24	.344	.404	.515
8	.621	.707	.834	26	.330	.388	.496
9	.582	.666	.798	28	.317	.374	.479
10	.549	.632	.765	30	.306	.361	.463
11	.521	.602	.735	40	.264	.312	.402
12	.497	.576	.708	50	.235	.279	.361
13	.476	.553	.684	60	.214	.254	.330
14	.457	.532	.661	80	.185	.220	.286
15	.411	.514	.641	100	.165	.196	.256
16	.426	.497	.623	250	.104	.124	.163
17	.412	.482	.606	500	.074	.088	.115
18	.400	.468	.590	1000	.050	.062	.081
19	.389	.456	.575				
20	.378	.444	.561	∞	0	0	0

−If $|r*| < r$, there is no basis to believe that there is an association between X and Y (i.e., $\varrho = 0$).

The following demonstrates the test.

Illustration 2

- For the above problem, we choose $\alpha = 0.05$.
- From Table 5-1, $r = 0.632$ for $n = 10$.
- Since $|r*| = 0.899 > r$, there is an association between oligomers and cure state at 95% confidence.

Repeating the analysis for the larger population sample, we obtain:

$$r = 0.299 \text{ (interpolating values in Table 5-1) for } n = 44$$

Again, since $|r*| = 0.46 > r$ there is an association.

CORRELATION FUNCTIONS AND SPECTRAL DENSITY ANALYSIS

A major objective in data analysis is to determine whether linear functions that are descriptive of data sets exist. These relations can often be extracted from a correlation function or its Fourier transform which is called the *spectral density function*. We shall now review the theory and mathematics of auto- and cross-correlation functions, and illustrate the usefulness of these definitions to random data analysis. An important point to keep in mind is that mathematics enables us to express or model raw data in a quantitative manner. Very often, regardless of the phenomenon or parameter under investigation, a suitable mathematical expression can be regressed. The mechanics of regression will be covered in the following chapter. For now, let's assume we can in fact represent our data in the form of an easily manipulated expression. In this regard, we consider the data as a continuous function or signal.

Consider two continuous random operations, $\{x(t)\}:\{y(t)\}$, which are ergodic and are separated by a time delay τ. Each can be represented by their individual time history records. The *covariance function* between the time history records $x(t)$ and $y(t)$ for a given time delay τ is:

$$C_{xy}(\tau) = E[\{x(t) - \mu_x\}\{y(t + \tau) - \mu_y\}]$$

$$= \lim_{T \to \infty} \frac{1}{T} \int_0^T \{x(t) - \mu_x\}\{y(t + \tau) - \mu_y\}dt \qquad (3)$$

$$= R_{xy}(\tau) - \mu_x\mu_y$$

where $R_{xy}(\tau)$ is the cross-correlation function between processes $x(t)$ and $y(t)$ and is defined as:

$$R_{xy}(\tau) = \lim_{T \to \infty} \frac{1}{T} \int_0^T x(t)y(t + \tau)dt \qquad (4)$$

There are situations where processes $x(t)$ and $y(t)$ are equal, in which case:

$$C_{xx}(\tau) = \lim_{T \to \infty} \frac{1}{T} \int_0^T \{x(t) - \mu_x\}\{x(t + \tau) - \mu_x\}dt = R_{xx}(\tau) - \mu_x^2 \qquad (5)$$

In this case R_{xx} is the autocorrelation function of $x(t)$ and is defined as:

$$R_{xx}(\tau) = \lim_{T \to \infty} \frac{1}{T} \int_0^T x(t)x(t + \tau)dt \qquad (6)$$

It can be shown that the value of the autocorrelation function at zero time lag is the mean square value of the data, i.e.:

$$R_{xx}(0) = \psi_x^2 = \sigma_x^2 + \mu_x^2 \qquad (7)$$

The autocorrelation function provides a measure of the degree of confidence in predicting conditions based on past experiments. It is used in conjunction with $p(x)$ to provide accurate predictions of $x(t)$ at the future time $t = T + \tau$. This can be illustrated by analyzing the signal of a simple repetitive function such as a sine wave. To do this, we need one more definition, that of the *spectral density function*.

The spectral density function is generally thought of in terms of the Fourier transform of a correlation function, but also has applications in generalized Fourier analysis. The spectral density function between the stationary random processes considered above may be expressed as:

$$S_{xy}(f) = \int_{-\infty}^{\infty} R_{xy}(\tau)e^{-j2\pi f\tau}d\tau \qquad (8)$$

When $x(t)$ and $y(t)$ are different, $S_{xy}(f)$ is called the cross-spectral density function; and when $y(t) = x(t)$ we obtain the autospectral density function $S_{xx}(f)$:

$$S_{xx}(f) = \int_{-\infty}^{\infty} R_{xx}(\tau)e^{-j2\pi f\tau}d\tau \qquad (9)$$

$S_{xx}(f)$ is also called the *power spectral density function*.

From the mathematics, spectral density functions are defined over all frequencies (both positive and negative); however, our concern most often lies with non-negative spectra. These are referred to as one-sided spectral density functions.

A useful relation to bear in mind is for the case where $\tau = 0$,

$$R_{xx}(o) = \int_{o}^{\infty} G_{xx}(f)df = \psi_x^2 = \sigma_x^2 + \mu_x^2 \qquad (10)$$

That is, the total area under the autospectral density function is simply the sum of the variance of the data and the square of the mean value of the data. It can be shown that:

$$S_{xx}(f) = \int_{-\infty}^{\infty} C_{xx}(\tau)e^{-j2\pi f\tau}d\tau + \mu_x^2\delta(f) \qquad (11)$$

(this is obtained by noting that $R_{xx}(\tau) = C_{xx}(\tau) + \mu_x^2$; where $\delta(f)$ is the delta function. The area under the autospectrum between any two frequency limits f_1 and f_2 yields the mean square value of the data over that frequency range:

$$\psi_x^2(f_1,f_2) = \int_{f_1}^{f_2} C_{xx}(f)df \qquad (12)$$

The cross-correlation function $R_{xx}(\tau)$ is defined by the inverse Fourier transform of the two-sided cross-spectrum $S_{xy}(f)$ [Equation (9)]:

$$R_{xy}(\tau) = \int_{-\infty}^{\infty} S_{xy}(f)e^{j2\pi f\tau}df \qquad (13)$$

And for a one-sided spectrum:

$$G_{xy}(f) = 2\int_{-\infty}^{\infty} R_{xy}(\tau)e^{-2j2\pi f\tau}d\tau = C_{xy}(f) - jQ_{xy}(f) \qquad (14)$$

$C_{xy}(f)$ is the real part of the function (called the coincident spectral density function or cospectrum) and $Q_{xy}(f)$ is the imaginary part (called the quadraspectrum). These definitions are as follows:

$$C_{xy}(f) = 2 \int_{-\infty}^{\infty} R_{xy}(\tau)\cos 2\pi f\tau d\tau \qquad (15)$$

$$Q_{xy}(f) = 2 \int_{-\infty}^{\infty} R_{xy}(\tau)\sin 2\pi f\tau d\tau \qquad (16)$$

Cross-spectra are often expressed in terms of a magnitude and phase angle:

$$G_{xy}(f) = | a_{xy}(f)|e^{-j\theta_{xy}(f)} \qquad (17)$$

where

$$|a_{xy}(f)| = \sqrt{C_{xy}^2(f) + Q_{xy}^2(f)}$$

$$\theta_{xy}(f) = \tan^{-1}\{Q_{xy}(f)/C_{xy}(f)\}$$

With the above relations defined we may now consider the sine wave signal show in Figure 5-4a.

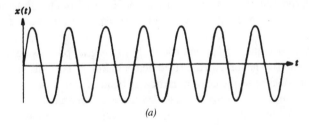

(a)

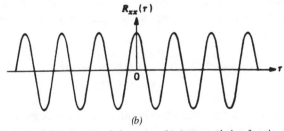

(b)

FIGURE 5-4. (a) Time history plot of sine wave. (b) Autocorrelation function of a sine wave.

The function can be written as:

$$x_k(t) = X\sin(2\pi f_o t + \theta_k) \tag{18}$$

where

$$p(\theta) = \tfrac{1}{2}\pi \quad , \quad 0 < \theta < 2\pi \tag{19}$$

The autocorrelation function of a sine wave is:

$$R_{xx}(\tau) = E[x_k(t)\, x_k(t + \tau)] \tag{20}$$

From this it follows that:

$$R_{xx}(\tau) = \frac{X^2}{2\pi} \int_{O}^{2\pi} \sin(2\pi f_o t + \theta)\, \sin\,[2\pi f_o(t + \tau) + \theta]d\theta \tag{21}$$

Performing the integration yields:

$$R_{xx}(\tau) = \frac{x^2}{2} \cos\, 2\pi f_o \tau \tag{22}$$

The autocorrelation function of a sine wave is therefore a cosine wave with an amplitude equivalent to the mean square value of the original signal. The autocorrelation function is shown in Figure 5-4b. Hence, from the time history of the signal over record length T, we are able to predict the sine wave at any time in the future.

The above concepts are applied routinely to a signal processing which can be divided into two general areas: (1) to detect unknown periodic signals in the presence of noise or to measure a particular band of signal or noise frequencies; (2) to compare two waveforms to determine the degree of conformity between them. In the latter application, both waveforms are often obscured by noise or other signals not common to each. The first case is the autocorrelation since it involves comparison between a waveform and a time-delayed version of itself. The function is expressed by Equation (6), or as follows:

$$R_{xx}(\tau) = \frac{1}{2T} \int_{-T}^{T} x_1(t)\, x_1(t - \tau)d\tau \tag{23}$$

Equation (23) allows for the extraction of information involving a signal which is buried in noise, but by the nature of the operation the output waveform may not be an actual reproduction of the input waveform. It will, however always be an even function. For example, a square wave can be expressed by the following Fourier series:

$$f(t) = \frac{4A}{\pi} (\cos \omega t - \frac{1}{3} \cos 3 \omega t + \frac{1}{5} \cos 5 \omega t - \ldots) \quad (24)$$

Upon substituting this expression into the autocorrelation function we obtain:

$$C_{11}(\tau) = \frac{8A^2}{\pi^2} (\cos \omega\tau + \frac{1}{3^2} \cos 3\omega\tau + \frac{1}{5^2} \cos 5\omega\tau + \ldots) \quad (25)$$

Equation (24) is the expression for a triangular wave. Hence, the autocorrelation of a square wave produces a triangular wave. The autocorrelation functions of a sinusoidal waveform have already been shown to be a cosine wave.

When noise is autocorrelated, the resulting waveform depends on the center frequency and bandwidth of the noise. The autocorrelation function for noise is:

$$C_{11}(\tau) = N e^{-B\tau} \cos \beta\tau \quad (26)$$

where N = the mean square value of noise, β = center of frequency, and B indicates the bandwidth.

If the expression for autocorrelation of a sine wave is combined with that of wide bandwidth noise, the resulting function is:

$$C_{11}(\tau) = N e^{-B\tau} + \frac{A^2}{2} \cos \omega\tau \quad (27)$$

The input and output waveforms for this example are shown in Figure 5-5. Examples of other waveforms and their corresponding autocorrelation functions are given in Figure 5-6.

The references at the end of the book provide detailed discussions of the principles of signal and random data analysis. The following illustration emphasizes some of the concepts presented.

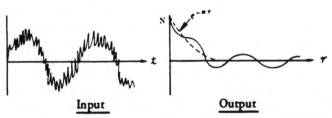

FIGURE 5-5. Combined waveforms for sine wave and wide bandwidth noise.

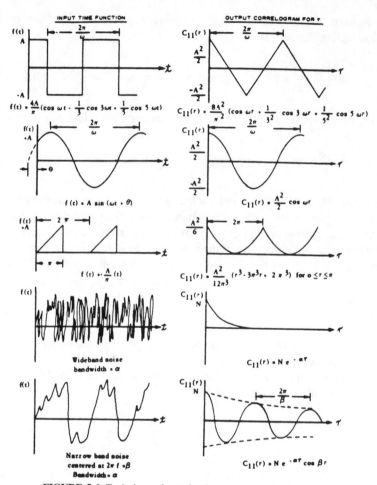

FIGURE 5-6. Typical waveforms for the autocorrelation function.

Illustration 3

A 2 ft. diameter fluidized bed reactor is showing poor yields. We believe ineffective gas-solids contacting may be the reason, particularly if the bed is operating in a slugging regime. The bed is normally operated at a superficial gas velocity of 0.42 fps. To assess the bed hydrodynamics we attempt to analyze the signals obtained from pressure transducers located at 10 ft. intervals along the height of the reactor. Oscillograph recordings of the pressure signals for different levels are shown in Figure 5-7. The recordings have been lined up under each other to show that they are the same, and are simply displaced in time. The triangular shape of individual signals may be indicative of gas slugs, where the break in the pressure rise portion of the signal corresponds to the gas slug reaching the pressure tap. If this interpretation is correct, then the time displacement between signals represents the mean time for gas slugs to travel between the transducers. Assuming that the slug does not break up or undergo extensive coalescence, the mean rise time for a gas slug is:

Base on Taps EF:

$$V_s = 10 \text{ ft}/4.2 \text{ sec} = 2.38 \text{ ft/s}$$

Based on Taps EG:

$$V_s = 20 \text{ ft}/8.4 \text{ sec} = 2.38 \text{ ft/s}$$

Based on Taps FG:

$$V_s = 10 \text{ ft}/(8.4 - 4.2 \text{ sec}) = 2.38 \text{ ft/sec}$$

Examining tap F for example, the average drop time, t, for each signal (i.e., the time for the signal to drop from peak to base as shown in Figure 5-7) is typically 0.6 ~ 1 sec ($\cong 0.8$ sec. mean). This piece of data along with the computed slug rise velocity enables us to estimate the length of the bubble:

$$D_b = 0.8 \text{ sec} \times 2.38 \text{ ft/sec} = 1.90 \text{ ft.}$$

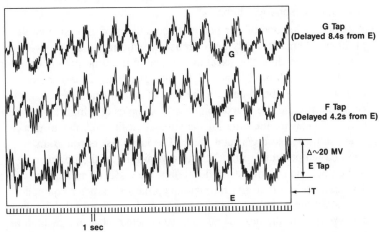

FIGURE 5-7. Oscillograph tracings of pressure transducer signals for Illustration 3.

Since the reactor is only 2 ft. in diameter, the analysis strongly suggests slugging conditions exist. As a check on the validity of the analysis, we should be able to material balance the gas flow. The total gas flow would be accounted for by summing the drop over a random time period. The signals in Figure 5-7 reveal a mean frequency spectrum on the order of 6 mv/sec. The transducers are calibrated for 0.05 kPa/mv and a typical pressure drop is 6 kPa for every 10 ft. of bed. Hence, the superficial gas velocity through the bed based on the analysis of the transducer signals is:

$$U_s = 6 \, \frac{mv}{sec} \times 0.05 \, \frac{kPa}{mv} \times \frac{10 \text{ ft. bed}}{6 \text{ kPa}} \times 8 \, \frac{\text{ft. solids}}{10 \text{ ft. bed}} = 0.40 \text{ fps.}$$

The last number (8 ft. solids/10 ft. bed) represents the bed voidage under minimum fluidization conditions (i.e., the actual amount of solids per 10 ft. of bed height at the point where the bed begins to fluidize). This calculated velocity agrees remarkably well with the 0.42 fps operating velocity.

PRACTICE PROBLEMS

1 Is there a correlation between the two extruder operations described in Illustration 8 (Chapter 4)? Estimate a correlation coefficient between the two operations.

2 Refer back to Practice Problem (2) in Chapter 3. (A) Provide an estimate of the mean bubble rise time. (B) Estimate the mean bubble size (assume spherical bubbles).

3 A customer for one of our polymers complains of excessive swell problems. Our chief polymer chemist hypothesizes that the swelling may be caused either by excessive crystallinity in the polymer and/or the effect of high molecular weight entanglements. To test these hypotheses, we obtain data on the heat of fusion (indicates crystallinity) and the relaxation time of samples at a high temperature (i.e., we measure the time for the stresses in the polymer to relax when subjected to a constant load). The latter test is run at an elevated temperature so that all the crystallinity of the polymer is melted out; hence, the test reflects only the effect of molecular weight. The samples are then run through a capillary rheometer equipped with a slit die and the polymer sample swell is measured. The data are reported on the next page. Can we conclude that one or both of these hypotheses are correct?

Polymer Sample	Heat of Fusion (cal/g)	Relaxation Time (sec.)	Swell (%)
A	0.42	85	17
B	0.004	67	10
C	0.003	84	10
D	0	100	15
E	0	98	15
F	0	93	10
G	0.042	39	28.1
H	0.039	50	24.5
I	NM	37	16.6
J	NM	31	17
K	NM	50	23
L	NM	29	24

NM – not measured

4 Plot the data for the following cases and make a guestimate of the correlation coefficient. Compute $r*$ for each case.

a) X	Y		b) X	Y		c) X	Y
50	50		10	-7		10	13
70	70		22	0.5		22	25
190	190		37	3.9		30	13

5 Can we state with 95% confidence that $\varrho \neq 0$ for the following:

$$a) \ r\bullet = 0.57, \quad n = 24$$
$$r* = 0.75, \quad n = 9$$
$$r* = 0.71, \quad n = 12$$
$$r* = 0.65, \quad n = 56$$

CHAPTER 6

Linear Regression and Interactive Models

INTRODUCTION

This chapter describes the mechanics of performing linear regressions. Linear regressions are equations that estimate the degree of linear relationship between two sets of variables. In addition, they indicate the equation of the line along which the variables are related. If a strong enough relationship exists between the variables, the linear equation generated by the regression can be used to predict the probable value of the dependent variable based on the known value of the independent variable. As is often the case, two or more factors contribute to the response variable, giving rise to the more complex situation of interaction. The effects of interaction parameters on regression models are also discussed and the mechanics of constructing a quadratic model are presented.

The present trend in industry is towards a greater use of personal computers. To this end, many corporations are investing in software that provides spreadsheet and word-processing capabilities. Many of the software spreadsheets on the market do not incorporate statistical regression capabilities. Discussions and illustrations are therefore provided for adapting spreadsheets to simple but widely used regression techniques. A list of commercial statistical software packages is also included.

Linear Regression

Before illustrating the method of least squares it is worth noting that the freehand method of curve fitting is useful in drawing an approximating curve or relationship to fit a set of data. If the type of equation for the curve is known, then the constants of the equation can be determined by selecting as many points on the curve as there are constants in the equation. For example, if the curve is linear, only two points are necessary; if it is a parabola, three points are necessary. The freehand method of curve fitting does have the disadvantage that the method requires individual judgment, and therefore different results will be obtained by different observers.

To avoid subjective curve fitting, the method of least squares described in this chapter can be used to arrive at the "best fitting curve." Figure 6-1 shows data points (X_1, Y_1), (X_2, Y_2), . . . (X_N, Y_N). For a given value of X (say X_o),

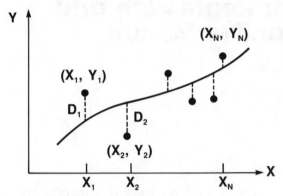

FIGURE 6-1. Method of least squares.

there will be a difference between the value Y_1 and corresponding values as determined by the curve. As shown in Figure 6-1, the difference (also called the *deviation, error* or *residual*) is denoted by D_1 and may be positive, negative or zero. Deviations for corresponding $X_2, \ldots, X_N$ values are $D_2, \ldots, D_N$.

An indication of how good the fit of the curve in Figure 6-1 is can be obtained from the sum of the squared deiviations, i.e., $D_1^2 + D_2^2 + \ldots + D_N^2$. When this value is small, the fit is good, and conversely, when it is large, the fit is poor. Hence, the objective of the method of least squares is to find the equation of a curve such that:

$$\min \{D_1^2 + D_2^2 + \ldots + D_N^2\}$$

The curve with this property is the *best fitting curve*. A line having this property is called a *least square line*.

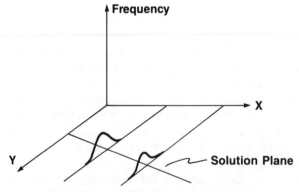

FIGURE 6-2. Conceptual approach to regression.

The general method used to arrive at the best fitting curve is referred to as linear regression analysis. The analysis estimates a function or model which will predict the behavior of a response variable Y over a range of factor levels X. We first treat the case of a single factor X, where a linear relationship between Y and X is sought. The relationship sought in the regression is illustrated conceptually in Figure 6-2, in which we can model the solution approach as follows:

$$Y_{ij} = \beta_o + \beta_1 X_1 + \epsilon_{ij} \tag{1}$$

where X_i = the i*th* factor level.

Y_{ij} = the response at the i*th* factor level and j*th* replication of the data.

ϵ_{ij} = the error at the i*th* factor level and j*th* replication.

β_o, β_1 = unknown parameters to be estimated from the regression (i.e., the intercept and slope of the linear relationship, respectively).

If no replicate experiments are run, then the j subscript is omitted.

The following assumptions are applied in the regression scheme:

1 X can be freely set to a value without error.

2 The error term ϵ is normally distributed with zero mean and variance σ^2 independent of the value of X.

Parameters β_o and β_1 are estimated by the least squares estimates b_o and b_1 to obtain the following linear estimation equation:

$$\hat{Y} = b_o + b_1 X \tag{2}$$

The estimates of b_o and b_1 are those values which minimize:

$$\min \sum_{i=1}^{n} (Y_i - \hat{Y}_i)^2 \tag{3}$$

The regression procedure is as follows:

- Compute the means $\bar{X}$ and $\bar{Y}$ and the deviations S_{xx}, S_{xy}, S_{yy} given in the last chapter. For convenience, these formulas are repeated below:

$$S_{xx} = \Sigma x^2 - (\Sigma x)^2/n \tag{4}$$

$$S_{yy} = \Sigma y^2 - (\Sigma y)^2/n \tag{5}$$

$$S_{xy} = \Sigma xy - (\Sigma x)(\Sigma y)/n \tag{6}$$

- Calculate the estimated slope of the linear equations, b_1 [for Equation (2)]:

$$b_1 = S_{xy}/S_{xx} \qquad (7)$$

- Compute the intercept estimate for Equation (2):

$$b_o = \bar{Y} - b_1\bar{X} \qquad (8)$$

- Calculate the variance estimate:

$$S^2 = \frac{S_{yy} - b_1^2 \, S_{xx}}{n - 2} \qquad (9)$$

The above procedure provides us with a least squares prediction equation that passes through the point $(\bar{Y}, \bar{X})$ and has a standard deviation of prediction S, and correlation coefficient r^*.

Illustration 1

A conductance probe is used to measure the liquid inventory in a holding tank. The probe has been calibrated by immersing it at different levels in the process fluid and measuring the output voltage. The calibration data, converted to gallons based on the tank's diameter are given below. Let's prepare a regression formula to obtain a direct reading of inventory during operation.

	X Output Voltage (mV)	Y Gallons
	0	0
	23	295
	36	405
	39	550
	67	850
	80	970
Total	245	3070
n	6	6
Average	40.8	511.7

$S_{xx} = 14{,}235 - (245)^2/6 = 4{,}230.8$

$S_{yy} = 2.217 \times 10^6 = (3070)^2/6 = 6.461 \times 10^5$

$S_{xy} = 1.774 \times 10^5 - (245)(3070)/6 = 52{,}007$

$b_1 = 52{,}007/4{,}230.8 = 12.29$

$b_o = 511.7 - 12.29(40.8) = 10.17$

$$S^2 = \frac{6.461 \times 10^5 - (12.29)^2(4{,}230.8)}{4} = 1765.7$$

$S = 42.0$

$r^* = S_{xy}/\sqrt{S_{xx}S_{yy}} = 52{,}007/\sqrt{(4230.8)(6.461 \times 10^5)} = 0.9947$

Hence, we have a regression equation: $\hat{Y} = 10.17 + 12.29X$ or

$$\text{Gallons} = 10.17 + 12.29 \text{ (Voltage)}$$

Although the correlation coefficient is high (i.e., a good fit) we have a standard deviation for prediction of 42 gallons. When the tank is near full capacity this amounts to less than 5% error, but at low inventories the predictive equation could be significantly off. For this reason, we need to quantify the confidence intervals of the regression model.

Confidence intervals can be established on β_1, $\beta_o + \beta_1 X_o$ and Y. They can be computed for the following:

- β_1, to test the hypothesis, $\mu_o : \beta_1 = 0$.
- On the expected value of Y at any given X (e.g., X_o). This is essentially the confidence limit on the model.
- On the future Y values at a given X (e.g., X_o). The procedure for calculating confidence intervals is as follows.
- Select the α-risk. Use the Student's t statistic (Table 2-2 in Chapter 2) for $n - 2$ degrees of freedom and $100 (1 - \alpha)\%$ confidence.
- Compute the confidence interval for β_1 from:

$$\beta_1 = b_1 \pm ts/\sqrt{S_{xx}} \tag{10}$$

- Compute the confidence interval on the error:

$$\epsilon(Y) = \beta_o + \beta_1 X \text{ at } X = X_o, \text{ using:}$$

$$\beta_o = b_o + b_1 X_o \pm ts \sqrt{\frac{1}{n} + \frac{(X_o - \bar{X})^2}{S_{xx}}} \tag{11}$$

- Compute the confidence interval on Y at X_o using:

$$Y = b_o + b_1X_o \pm ts \sqrt{1 + \frac{1}{n} + \frac{(X_o - \bar{X})^2}{S_{xx}}} \qquad (12)$$

Illustration 2

Continuing from Illustration 1, we have:

$\nu = 6 - 2 = 4$ and we choose $\alpha = 0.05$. Hence, from Table 2-2 in Chapter 2, $t = 2.776$.

For the β_1 confidence interval, use Equation (10):

$$12.29 \pm 2.776 \, (42)/\sqrt{4230.8}$$

$$12.29 \pm 1.79$$

$$\therefore \qquad 10.50 < \beta_1 < 14.08$$

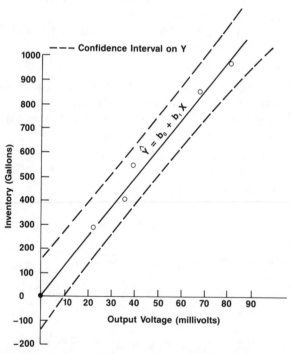

FIGURE 6-3. Shows confidence intervals on predicted Y for Illustration 2.

Next, at $X_o = 0$, we compute the confidence interval on β_o using Equation (11):

$$10.17 + 12.29(0) \pm 2.776(42) \sqrt{\frac{1}{6} + \frac{(0 - 40.8)^2}{4230.8}}$$

$$10.77 \pm 87.26$$

$$\therefore \quad -77.1 \leq \beta_o \leq 97.43$$

Finally, for the confidence interval on Y at X_o, use Equation (12):

$$10.17 + 12.29(0) \pm 2.776(42) \sqrt{1 + \frac{1}{6} + \frac{(0 - 40.8)^2}{4230.8}}$$

$$10.17 \pm 145.6$$

$$\therefore \quad -135.4 < Y < 155.8$$

In the above example, the calculations were illustrated for $X_o = 0$. We can get a better feel for the confidence interval over the entire range of the correlation by choosing X_o differently. Figure 6-3 shows the predictive Y confidence limit graphically over the entire range of the regression formula.

The method of least squares can also be applied to non-linear equations which are reducible to linear form. The following examples illustrate such applications.

Illustration 3

The following viscosity data were obtained for a polymer in a capillary rheometer:

Shear rate, $\gamma(\text{sec}^{-1})$	2.5	10	35	170	550	1800
Viscosity, $\mu(\text{kPa-s})$	170	88	39	13	5.2	2.6

Viscosity data are normally correlated with a power law fit: $\mu = k\gamma^{n'}$, where k and n' are constants.

We shall apply the method of least squares by rewriting this expression in a linear form as follows:

$$\log\mu = \log k + n'\log\gamma$$

In terms of the standard expression for a straight line [Equation (12)], we let

$\log\dot{\gamma} = X$, $log\mu = Y$ and $b_o = \log k$. Hence, our expression is:

$$Y = b_o + b_1 X$$

The following tabulations summarize the calculations for obtaining the least square line.

$X = \log\dot{\gamma}$	$Y = \log\mu$	X^2	Y^2	XY
0.3979	2.2304	0.1583	4.9747	0.8875
1.0000	1.9445	1.0000	3.7811	1.9445
1.5441	1.5911	2.3842	2.5316	2.4568
2.2304	1.1139	4.9747	1.2408	2.4844
2.7404	0.7160	7.5098	0.5127	1.9621
3.2553	0.4150	10.5970	0.1722	1.3509

$\Sigma X = 11.1681$ $\Sigma Y = 8.0109$ $\Sigma X^2 = 26.6240$ $\Sigma Y^2 = 13.2131$ $\Sigma XY = 11.0862$

$\bar{X} = 1.8614$ $\bar{Y} = 1.3352$

$$S_{xx} = 26.6240 - (11.1681)^2/6 = 5.8363$$

$$S_{yy} = 13.2131 - (8.0109)^2/6 = 2.5174$$

$$S_{xy} = 11.0862 - (11.1681)(8.0109)/6 = -3.8249$$

$$b_1 = -3.8249/5.8363 = -0.6554$$

$$b_o = 1.3352 - (-0.6554)(1.8614) = 2.5552$$

$$Y = 2.552 - 0.6554X$$

Since $b_o = \log k$, then $k = 10$, $b_o = 359.1$ and $b_1 = n' = -0.6554$. Hence the power law expression is:

$$\mu = 359.1\dot{\gamma}^{-0.6554}$$

The correlation coefficient for the linear regression is:

$$|r^*| = \left| \frac{-3.8249}{\sqrt{5.8363 \times 2.5174}} \right| = 0.9979$$

To obtain a better feel for how good the correlation is, we can compare the power law expression predictions to the measurements:

$\dot{\gamma}$	Measured μ	Computed μ	Deviation
2.5	170	196.9	26.9
10	88	79.4	−8.6
35	39	34.9	−4.1
170	13	12.4	−0.6
550	5.2	5.7	0.5
1800	2.6	2.6	0.0

The reader may wish to compare the variance estimate to complete the regression.

Illustration 4

We shall develop an exponential regression formula for the cure state data described in Illustration 1 of Chapter 5. For the sake of brevity, we shall randomly select about 50% of the data reported in Figure 5-3b. The form of the expression we wish to develop is:

$$MH = \alpha e^{\beta Olig}$$

where MH = cure state
Olig = Oligomers content
α, β = constants

This can be reduced to a linear expression:

$$Y = b_o + b_1 X$$

by letting $y = 1n(MH)$
$b_o = 1n(\alpha)$
X = Olig.
$b_1 = \beta$

The necessary computations are tabulated in Table 6-1, from which we may compute the following statistics coefficients:

$$S_{xx} = 4.5759 \times 10^5 - (2.449 \times 10^3)^2/22 = 1.8497 \times 10^5$$

$$S_{yy} = 304.78 - (81.87)^2/22 = 0.1196$$

$$S_{xy} = 9.0206 \times 10^3 - (2.449 \times 10^3)(81.87)/22 = -93.02$$

Table 6-1. Summary of Calculations for Illustration 4.

X Oligomers	MH	y = ln(MH)	x^2	y^2	xy
68	43	3.761	4.624×10^3	14.145	255.75
71	41.7	3.731	5.041	13.920	264.90
49	42	3.738	2.401	13.973	183.16
66	42.4	3.747	4.356	14.040	247.30
76	43.2	3.766	5.776	14.183	286.22
37	43.1	3.764	1.369	14.168	139.27
73	41.9	3.735	5.329	13.950	272.66
74	43.4	3.770	5.476	14.213	278.98
66	41.5	3.726	4.356	13.883	245.92
70	41.3	3.721	4.900	13.846	260.47
130	40.2	3.694	16.900	13.646	480.22
101	41.2	3.718	10.201	13.823	375.52
144	39.0	3.664	20.736	13.425	527.62
152	39.1	3.666	23.104	13.439	557.23
199	37.6	3.627	39.601	13.155	721.77
187	38.9	3.667	34.969	13.403	684.61
304	40.3	3.696	92.416	13.660	1123.58
397	36.2	3.589	157.609	12.881	1424.83
0.1	45.1	3.809	1×10^{-5}	14.508	0.381
0.1	46.4	3.837	1×10^{-5}	14.723	0.384
67.1	41.6	3.728	4.502	13.898	250.19
11.8	41.5	3.726	13.924	13.883	439.67
$\Sigma X = 2.449 \times 10^3$		$\Sigma Y = 81.87$	$\Sigma X^2 = 4.5759 \times 10^5$	$\Sigma y^2 = 304.478$	$\Sigma xy = 9.0206 \times 10^3$

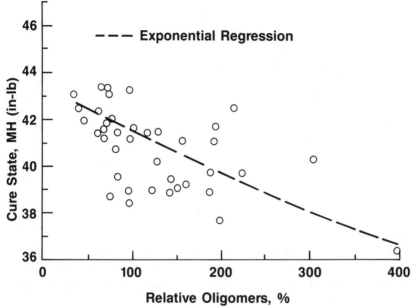

FIGURE 6-4. Exponential regression plot for Illustration 4.

$b_1 = S_{xy}/S_{xx} = -93.02/1.8497 \times 10^5 = -5.029 \times 10^{-4}$

$b_o = \bar{Y} - b_1\bar{X} = 3.721 - (-5.029 \times 10^{-4})(111.32)$

$b_o = 3.777$

Hence, the linear expression is: $Y = 3.777 - 5.029 \times 10^{-4}X$, or after being converted to the desired form of the expression:

$$MH = 43.68 \exp(-5.029 \times 10^{-4} \text{ Olig.}).$$

The correlation is shown plotted along with the measurements in Figure 6-4.

Quadratic Models

We have shown that the response of an experimental design can be expressed as a mathematical function of the factors. Often, empirical correlations based on first-, second-, or third-order polynomials form the basis for a system or parameter model. The underlying assumption in the usefulness of such empirical models is that both the responses and the factors can be treated as continuous functions.

The obvious advantage of a linear model is that it is simple to use; however,

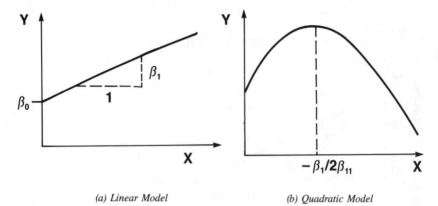

(a) Linear Model   (b) Quadratic Model

FIGURE 6-5. Compares linear and quadratic models.

nature is much more complex and, therefore, is better approximated by non-linear expressions. Linear models are, however, quite adequate for identifying the existence of a factor effect and, hence, are most useful during the initial stages of an investigation.

We now turn our attention to the quadratic model which provides us an opportunity to identify an approximate optimum. Both the linear and quadratic models are illustrated graphically in Figure 6-5, and the quadratic model takes the form of an expression as follows:

$$Y = \beta_o + \beta_1 X + \beta_{11} X^2 \tag{13}$$

The β coefficients are the model parameters, or more specifically:

$$\beta_o = \text{intercept}$$

$$\beta_1 = \text{slope}$$

$$\beta_{11} = \text{curvature}$$

The coefficients are found from the normal equations:

$$\Sigma Y = \beta_o N + \beta_1 X + \beta_{11} X^2$$

$$\Sigma XY = \beta_o \Sigma X + \beta_1 \Sigma X^2 + \beta_2 \Sigma X^3 \tag{14}$$

$$\Sigma X^2 Y = \beta_o \Sigma X^2 + \beta_1 \Sigma X^3 + \beta_2 \Sigma X^4$$

The following example illustrates this least squares parabola.

Illustration 5

Data on the specific heat of carbon dioxide as a function of temperature are given below. We shall prepare a quadratic model.

Tempera- ture, $T(°F)$	0	30	60	80	100	150	200	300	400	1000
Specific Heat, C_p (Btu/lb-F)	0.193	0.198	0.202	0.204	0.207	0.213	0.219	0.230	0.239	0.280

To make calculations simpler in terms of significant digits, we shall normalize the temperature scale by dividing through by 100; i.e., $X = T/100$, $y = C_p$.

i	X	Y	X^2	X^3	X^4	XY	X^2Y
1	0	0.193	0	0	0	0	0
2	0.3	0.198	0.090	0.027	8.100×10^{-3}	5.94×10^{-2}	1.782×10^{-2}
3	0.6	0.202	0.360	0.216	0.1296	0.1212	7.272×10^{-2}
4	0.8	0.204	0.640	0.512	0.4096	0.1632	0.1306
5	1.0	0.207	1.000	1.000	1.000	0.2070	0.2070
6	1.5	0.213	2.250	3.375	5.063	0.3195	0.4793
7	2.0	0.219	4.000	8.000	16.000	0.4380	0.8760
8	3.0	0.230	9.000	27.000	81.000	0.6900	2.0700
9	4.0	0.239	16.000	64.000	256.000	0.9560	3.8240
10	10.0	0.280	100.000	1000.000	1.000×10^4	2.800	28.000

$\Sigma X=23.2$ $\Sigma Y=2.185$ $\Sigma X^2=133.3$ $\Sigma X^3=1.104 \times 10^3$ $\Sigma X^4=1.036 \times 10^4$ $\Sigma XY=5.754$ $\Sigma X^2Y=35.677$

Preparing Equation set (14), we obtain:

$$2.185 = 10\beta_o + 23.2\beta_1 + 133.3\beta_{11}$$

$$5.754 = 23.2\beta_o + 133.3\beta_1 + 1.104 \times 10^3\beta_{11}$$

$$35.677 = 133.3\beta_o + 1.104 \times 10^3\beta_1 + 1.036 \times 10^4\beta_{11}$$

Hence, we have three equations with three unknowns which can be solved for simultaneously. The coefficients are:

$$\beta_o = 0.2000, \qquad \beta_1 = 8.069 \times 10^{-3}, \qquad \beta_{11} = -1.791 \times 10^{-5}$$

Hence, our quadratic equation is:

$$C_p = 0.2000 + 8.069 \times 10^{-3}(T/100) - 1.791 \times 10^{-5}(T/100)^2$$

The trend values predicted from the model are compared to the measurements below:

$T(°F)$	C_p(meas.)	C_p(trend)	Deviation
0	0.193	0.200	0.007
30	0.198	0.202	0.004
60	0.202	0.205	0.003
80	0.204	0.206	0.002
100	0.207	0.208	0.001
150	0.213	0.212	−0.001
200	0.219	0.216	−0.003
300	0.230	0.224	−0.006
400	0.239	0.232	−0.007
1000	0.280	0.279	−0.001

$$\text{Deviations Squared} = 1.75 \times 10^{-4}$$

Multiple Regression

Detailed discussions on multiple correlation and regression techniques are beyond the scope of this short reference manual; nevertheless, we shall try to highlight some useful concepts and guidelines. Whenever two or more factors influence a response, the situation referred to as *interaction* may exist. Interaction refers to the case where the response is not additive with respect to the two factors (refer to Figure 6-6). The linear model that reflects interaction for two factors is:

$$Y = \beta_o + \beta_1 X_1 + \beta_2 X_2 + \beta_{12} X_1 X_2 \tag{15}$$

where the coefficient β_{12} is defined as the *interaction parameter*. Equation (15) can be manipulated algebraically into a more instructive form:

$$Y = (\beta_o + \beta_2 X_2) + (\beta_1 + \beta_{12} X_2) X_1 \tag{16}$$

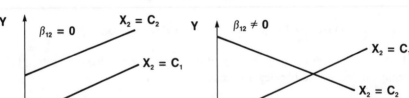

FIGURE 6-6. Shows difference between additive and interactive models.

With this alternate relation we have expressed response Y in terms of factor X_1, and X_2 is treated like a fixed constant. For the case of $\beta_{12} = 0$, the effect of X_2 is that it changes the intercept of the Y-Y_1 model. If $\beta_{12} \neq 0$, then the effect of X_2 becomes apparent in the slope of the Y-X_1 model as well as the intercept.

A quadratic model is generally flexible enough to simulate most non-linear response functions. The full quadratic model is written as:

$$Y = \beta_o + \beta_1 X_1 + \beta_2 X_2 + \beta_{11} X_1^2 + \beta_{12} X_1 X_2 + \beta_{22} X_2^2 \quad (17)$$

This type of relation often provides a good correlation over the range of experimental data, however, it should not be used for extrapolation much beyond the measurement region.

With three or more factors, the general second order model for n factors takes the form:

$$Y = \beta_o + \sum_{i=1}^{n} \beta_i X_i + \sum_{i=1}^{n} \sum_{j=1}^{n} \beta_{ij} X_i X_j \quad (18)$$

With this model there are n first-order terms (i.e., slopes) and $n(n - 1)/2$ interaction terms, and n quadratic terms.

The fundamental principles applied to problems of multiple correlation and regression are analogous to those of simple correlation treated earlier in this chapter. To generalize for large numbers of variables let's first adopt some standard notation. Let $X_1, X_2, X_3, \ldots$ denote values assumed by the variable X_1 (and $X_{21}, X_{22}, X_{23}, \ldots$ denote values assumed by the variable X_2, etc.). We can therefore express sums as:

$$X_{21} + X_{22} + X_{23} + \cdots + X_{2N} = \sum_{j=1}^{N} X_{2j} = \Sigma X_2 \quad (19)$$

As explained in this chapter, a regression equation is a model for estimating some dependent variable (e.g., X_1) from the independent variables X_2, $X_3, \ldots$. This is the regression formula of X_1 on $X_2, X_3, \ldots$. Many mathematics textbooks express this in terms of functional notation:

$$X_1 = F(X_2, X_3, \ldots) \quad (20)$$

In this expression, $Y = X_1$ is adopted in order to utilize proper subscript notation. Let's assume that we can model a three-variable case with a very simple regression formula having the form:

$$X_1 = \beta_{1.23} + \beta_{12.3} X_2 + \beta_{13.2} X_3 \quad (21)$$

$\beta_{1.23}, \beta_{12.3}, \beta_{13.2} = $ constants.

The subscripts after the dot indicate the variable that is maintained constant. For example, if X_3 is kept constant, then a plot of X_1 vs. X_2 would result in a straight line with slope $\beta_{12.3}$, as we have explained earlier. Again, if we keep X_2 constant and graph X_1 vs. X_3, we obtain a straight line with slope $\beta_{13.2}$. Note however that X_1 will vary partially because of variation in X_2 and partially because of variation in X_3. For this reason, $\beta_{12.3}$ and $\beta_{13.3}$ are more correctly referred to as the *partial regression coefficients* of X_1 on X_2 keeping X_3 constant, and X_1 on X_3 keeping X_2 constant, respectively.

Equation (21) is simply a linear regression equation of X_1 (or Y) on X_2 and X_3. Up to now, we have represented such correlations on a two-dimensional scatter plot. In a three-dimensional rectangular coordinate system we essentially construct a regression plane (refer back to Figure 6-1). This means that there exist least squares regression planes that fit N data points (X_1, X_2, X_3) on a three-dimensional coordinate system.

A regression formula of the form of Equation (21) can be derived by solving simultaneously the *normal equations*:

$$\left.\begin{aligned}
\Sigma X_1 &= \beta_{1.23}N + \beta_{12.3}\Sigma X_2 + \beta_{13.2}\Sigma X_3 \\
\Sigma X_1 X_2 &= \beta_{1.23}\Sigma X_2 + \beta_{12.3}\Sigma X_2^2 + \beta_{13.2}\Sigma X_2 X_3 \\
\Sigma X_1 X_3 &= \beta_{1.23}\Sigma X_3 + \beta_{12.3}\Sigma X_2 X_3 + \beta_{13.2}\Sigma X_3^2
\end{aligned}\right\} \tag{22}$$

The expressions are obtained by multiplying both sides of Equation (21) by 1, X_2 and X_3 successively and summing on both sides.

To simplify calculations, the variables can be coded in the following manner:

$$\chi_1 = X_1 - \bar{X}_1,$$

$$\chi_2 = X_2 - \bar{X}_2,$$

$$\chi_3 = X_3 - \bar{X}_3$$

In this manner, the regression equation of X_1 on X_2 and X_3 can be written as:

$$\chi_1 = \beta_{12.3}\chi_2 + \beta_{13.2}\chi_3 \tag{23}$$

The partial regression coefficients $\beta_{12.3}$ and $\beta_{13.2}$ can be obtained by solving the following two equations simultaneously:

$$\left.\begin{aligned}
\Sigma \chi_1 \chi_2 &= \beta_{12.3}\Sigma \chi_2^2 + \beta_{13.2}\Sigma \chi_2 \chi_3 \\
\Sigma \chi_1 \chi_3 &= \beta_{12.3}\Sigma \chi_2 \chi_3 + \beta_{13.2}\Sigma \chi_3^2
\end{aligned}\right\} \tag{24}$$

These equations are equivalent to the normal Equations (22).

The following example illustrates a three-variable linear regression along with estimation of the standard deviation and correlation coefficient.

Illustration 6

There may be several variables which would effect the cure state properties of the terpolymer described in Illustration 4. Two other variables are the diene content of the polymer and the Mooney viscosity. We can assess the effect of these variables separately on the cure state, MH, using a three variable model. In this example X_1 = MH, X_2 = Oligomers and X_3 = diene content or Mooney viscosity. We may compute the standard deviation of variable X_1, X_2 and X_3 using the following formulas:

$$S_1 = \sqrt{\frac{\Sigma X_1^2}{N} - \left(\frac{\Sigma X_1}{N} \right)^2}$$

$$S_2 = \sqrt{\frac{\Sigma X_2^2}{N} - \left(\frac{\Sigma X_2}{N} \right)^2} \tag{25}$$

$$S_3 = \sqrt{\frac{\Sigma X_3^2}{N} - \left(\frac{\Sigma X_3}{N} \right)^2}$$

Also, the correlation coefficient can be computed as a linear coefficient between X_1 and X_2, ignoring X_3:

$$r_{12} = \frac{N\Sigma X_1 X_2 - (\Sigma X_1)(\Sigma X_2)}{\sqrt{[N\Sigma X_1^2 - (\Sigma X_1)^2]\,[N\Sigma X_2^2 - (\Sigma X_2)^2]}} \tag{26}$$

An analogous expression can be written for the coefficient between X_1 and X_3, ignoring X_2.

Table 6-2 shows the necessary computations performed using a standard spreadsheet on a PC for Mooney viscosity as the third variable. Similar calculations were carried out for diene and a summary of the standard deviations and correlation coefficients for the two cases is as follows:

X_3 Variable	S_1	S_2	S_3	r_{12}^*	r_{13}^*
Diene	2.157	5.982	0.483	0.414	0.121
Mooney	2.157	5.982	7.687	0.414	0.605

Although the variance for $X_1 - X_3$ is much greater for the Mooney case than for diene, a much higher correlation is obtained for Mooney. To test for whether $\varrho \neq 0$, we refer to Table 5-1 in Chapter 5. For $\alpha = 0.05$ and $n = 38$ $r \cong 0.31$. For diene, $r_{13}^* < r$, hence no correlation and $\varrho = 0$; but for the

Table 6-2. Summary of Calculations for Illustration 6.

sample no.	X1 MH (in-lb)	X2 Olig-omers	X3 ML(1+8) @ 127C	X1^2	X2^2	X3^2	X1*X2	X1*X3	X2*X3
1.0000E+00	4.1300E+01	7.0000E+01	6.7000E+01	1.7057E+03	4.9000E+03	4.4890E+03	2.8910E+03	2.7671E+03	4.6900E+03
2.0000E+00	3.9550E+01	8.5000E+01	7.0500E+01	1.5642E+03	7.2250E+03	4.9703E+03	3.3618E+03	2.7883E+03	5.9925E+03
3.0000E+00	3.9000E+01	1.2400E+02	6.4000E+01	1.5210E+03	1.5376E+04	4.0960E+03	4.8360E+03	2.4960E+03	7.9360E+03
4.0000E+00	4.0700E+01	8.1000E+01	7.0000E+01	1.6565E+03	6.5610E+03	4.9000E+03	3.2967E+03	2.8490E+03	5.6700E+03
5.0000E+00	3.8950E+01	1.4400E+02	6.6900E+01	1.5171E+03	2.0736E+04	4.4756E+03	5.6088E+03	2.6058E+03	9.6336E+03
6.0000E+00	3.9100E+01	1.5200E+02	7.0000E+01	1.5288E+03	2.3104E+04	4.9000E+03	5.9432E+03	2.7370E+03	1.0640E+04
7.0000E+00	3.7600E+01	1.9900E+02	7.1500E+01	1.4138E+03	3.9601E+04	5.1123E+03	7.4824E+03	2.6884E+03	1.4229E+04
8.0000E+00	4.1500E+01	8.7000E+01	8.0000E+01	1.7223E+03	7.5690E+03	6.4000E+03	3.6105E+03	3.3200E+03	6.9600E+03
9.0000E+00	4.4400E+01	1.3100E+02	8.0500E+01	1.9714E+03	1.7161E+04	6.4803E+03	5.8164E+03	3.5742E+03	1.0546E+04
1.0000E+01	4.3900E+01	2.3800E+02	7.6000E+01	1.9272E+03	5.6644E+04	5.7760E+03	1.0448E+04	3.3364E+03	1.8088E+04
1.1000E+01	3.9500E+01	1.4400E+02	6.8500E+01	1.5603E+03	2.0736E+04	4.6923E+03	5.6880E+03	2.7058E+03	9.8640E+03
1.2000E+01	4.1700E+01	1.9600E+02	6.6000E+01	1.7389E+03	3.8416E+04	4.3560E+03	8.1732E+03	2.7522E+03	1.2936E+04
1.3000E+01	4.1050E+01	1.9300E+02	6.5000E+01	1.6851E+03	3.7249E+04	4.2250E+03	7.9227E+03	2.6683E+03	1.2545E+04
1.4000E+01	4.1500E+01	1.1800E+02	7.2500E+01	1.7223E+03	1.3924E+04	5.2563E+03	4.8970E+03	3.0088E+03	8.5550E+03
1.5000E+01	3.8900E+01	1.8700E+02	7.5000E+01	1.5132E+03	3.4969E+04	5.6250E+03	7.2743E+03	2.9175E+03	1.4025E+04
1.6000E+01	3.9700E+01	2.2500E+02	7.0000E+01	1.5761E+03	5.0625E+04	4.9000E+03	8.9325E+03	2.7790E+03	1.5750E+04
1.7000E+01	3.9700E+01	1.8900E+02	7.7000E+01	1.5761E+03	3.5721E+04	5.9290E+03	7.5033E+03	3.0569E+03	1.4553E+04
1.8000E+01	3.9250E+01	1.6200E+02	7.3000E+01	1.5406E+03	2.6244E+04	5.3290E+03	6.3585E+03	2.8653E+03	1.1826E+04
1.9000E+01	4.1650E+01	3.1240E+02	7.2000E+01	1.7347E+03	9.7594E+04	5.1840E+03	1.3011E+04	2.9988E+03	2.2493E+04
2.0000E+01	4.1100E+01	3.5030E+02	7.6000E+01	1.6892E+03	1.2271E+05	5.7760E+03	1.4397E+04	3.1236E+03	2.6623E+04
2.1000E+01	4.3350E+01	9.8300E+01	5.9000E+01	1.8792E+03	9.6629E+03	3.4810E+03	4.2613E+03	2.5577E+03	5.7997E+03

(continued)

Table 6-2. Summary of Calculations for Illustration 6. (continued).

sample no.	X1 MH (in-lb)	X2 Olig- omers	X3 ML(1+8) @ 127C	X1^2	X2^2	X3^2	X1*X2	X1*X3	X2*X3
2.2000E+01	4.0250E+01	3.0430E+02	6.0000E+01	1.6201E+03	9.2598E+04	3.6000E+03	1.2248E+04	2.4150E+03	1.8258E+04
2.3000E+01	4.2500E+01	2.1780E+02	7.1000E+01	1.8063E+03	4.7437E+04	5.0410E+03	9.2565E+03	3.0175E+03	1.5464E+04
2.4000E+01	3.6200E+01	3.9740E+02	5.6000E+01	1.3104E+03	1.5793E+05	3.1360E+03	1.4386E+04	2.0272E+03	2.2254E+04
2.5000E+01	4.5050E+01	0.0000E+00	8.2000E+01	2.0295E+03	0.0000E+00	6.7240E+03	0.0000E+00	3.6941E+03	0.0000E+00
2.6000E+01	4.6350E+01	0.0000E+00	9.3500E+01	2.1483E+03	0.0000E+00	8.7423E+03	0.0000E+00	4.3337E+03	0.0000E+00
2.7000E+01	4.1550E+01	6.7100E+01	6.9000E+01	1.7264E+03	4.5024E+03	4.7610E+03	2.7880E+03	2.8670E+03	4.6299E+03
2.8000E+01	4.3850E+01	6.6700E+01	7.0000E+01	1.9228E+03	4.4489E+03	4.9000E+03	2.9248E+03	3.0695E+03	4.6690E+03
2.9000E+01	4.4000E+01	7.1400E+01	8.6000E+01	1.9360E+03	5.0980E+03	7.3960E+03	3.1416E+03	3.7840E+03	6.1404E+03
3.0000E+01	4.3550E+01	8.3600E+01	7.7500E+01	1.8966E+03	6.9890E+03	6.0063E+03	3.6408E+03	3.3751E+03	6.4790E+03
3.1000E+01	4.1500E+01	1.3050E+02	7.3000E+01	1.7223E+03	1.7030E+04	5.3290E+03	5.4158E+03	3.0295E+03	9.5265E+03
3.2000E+01	4.1100E+01	1.5910E+02	6.0000E+01	1.6892E+03	2.5313E+04	3.6000E+03	6.5390E+03	2.4660E+03	9.5460E+03
3.3000E+01	3.8750E+01	7.6000E+01	6.3000E+01	1.5016E+03	5.7760E+03	3.9690E+03	2.9450E+03	2.4413E+03	4.7880E+03
3.4000E+01	3.8500E+01	9.8300E+01	5.7500E+01	1.4823E+03	9.6629E+03	3.3063E+03	3.7846E+03	2.2138E+03	5.6523E+03
3.5000E+01	3.9000E+01	9.7400E+01	7.0500E+01	1.5210E+03	9.4868E+03	4.9703E+03	3.7986E+03	2.7495E+03	6.8667E+03
3.6000E+01	4.0200E+01	1.2950E+02	7.4000E+01	1.6160E+03	1.6770E+04	5.4760E+03	5.2059E+03	2.9748E+03	9.5830E+03
3.7000E+01	4.1200E+01	1.0100E+02	6.5500E+01	1.6974E+03	1.0201E+04	4.2903E+03	4.1612E+03	2.6986E+03	6.6155E+03
3.8000E+01	4.1000E+01	7.9300E+01	6.7000E+01	1.6810E+03	6.2885E+03	4.4890E+03	3.2513E+03	2.7470E+03	5.3131E+03
	1.5580E+03	5.5654E+03	2.6859E+03	6.4051E+04	1.1063E+06	1.9209E+05	2.2520E+05	1.1050E+05	3.8514E+05
	X1	X2	X3	X1^2	X2^2	X3^2	X1*X2	X1*X3	X2*X3
	S2	S3	r12	r13					
S1 2.1566E+00	5.9823E+00	7.6871E+00	-4.1440E-01	6.0467E-01					

Mooney r_{13}^* > r and we can conclude that $\varrho \neq 0$. In fact, the reason why diene seems not to have an effect is that the range of this factor is quite small. We therefore can develop a regression equation by first writing the normal equations [Equation (22)] and then solving for the partial correlation coefficients simultaneously.

$$1.558 \times 10^3 = 38\beta_{1.23} + 5.565 \times 10^3\beta_{12.3} + 2.686 \times 10^3\beta_{13.2}$$

$$2.252 \times 10^5 = 5.565 \times 10^3\beta_{1.23} + 1.1063 \times 10^6\beta_{12.3} + 3.851 \times 10^5\beta_{13.2}$$

$$1.105 \times 10^5 = 2.686 \times 10^3\beta_{1.23} + 3.851 \times 10^5\beta_{12.3} + 1.9209 \times 10^5\beta_{13.2}$$

The final regression formula is:

$$X_1 = 42.72 - 0.01207X_2 + 7.239 \times 10^{-4}X_3$$

A parity plot of the model predictions of MH versus measured values is shown in Figure 6-7. Although the correlation is not a very good fit, it does provide a rough guide by which we can gauge the effect of Mooney and oligomers. From the standpoint of product quality, we would make a polymer over a narrow Mooney range. As shown by the correlation coefficient on X_3, the effect

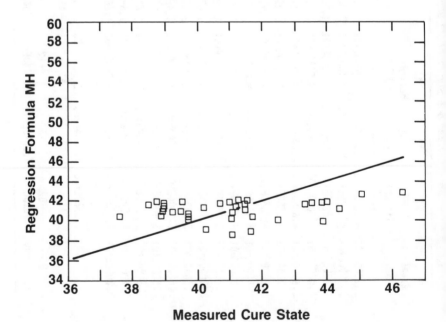

FIGURE 6-7. Parity plot for Illustration 6.

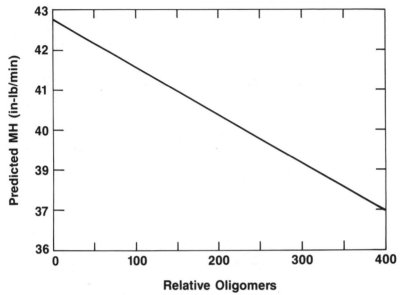

FIGURE 6-8. Shows effect of oligomers on MH from predictions.

of Mooney is quite small and will therefore not impact on cure properties once an acceptable range has been selected (usually production is 10 point standard deviation on Mooney). Oligomers however show a steep slope (partial correlation coefficient as shown in Figure 6-8). In this particular problem, the oligomers were generated by the presence of HCl during polymerization which gave rise to cationic reactions producing low molecular weight species (oligomers). The low molecular weight species have a detrimental effect on cure properties. By using an appropriate acid scavenger to remove the HCl, better cure properties in the product were obtained.

Performing Regressions on Spreadsheets

The use of spreadsheets on PC's (personal computers) has become popular in recent years. Spreadsheet software provides a convenient format for archiving data and performing simple manipulations. In fact, most spreadsheet software systems have simple graphics routines so that scatter plots can be instantly generated. In addition, software such as Lotus Development Corporation's 1-2-3 can rank data sets in order of preference.

The disadvantage with commercial spreadsheet software is that they are not designed to perform regressions on their own. Often a second software package specifically designed for statistical techniques such as regressions must be purchased. This second-generation software may have to be run independently of the spreadsheet, although more software vendors are preparing spreadsheet

Table 6-3a. Commercial Statistical Software Packages.

Product	Vendor	Minimum memory required (in K bytes)	Hard disk required	Programs integrated	Batch	Interactive	Both	ASCII	DIF	dBASE II	Lotus worksheet	Frequencies	Percentages	Cross tabulation	Factor analysis	Analysis of variance	Multivariate analysis	Linear	Non-linear	Multiple	Time series analysis	High resolution	Histograms	Scatter plots	Line graphs	Bar graphs	Report writing	Data management	Sort files	Join files
Simulations	Actuarial Micro Software	128K		✓		✓		✓	✓	✓		✓	✓			✓	✓	✓	✓	✓			✓		✓	✓	✓	✓	✓	
MODLER BLUE	Alphametrics Corp.	512K	✓	✓			✓	✓	✓		✓	✓	✓			✓	✓	✓		✓		✓	✓	✓	✓	✓		✓		✓
MODLER 100[1]		640K	✓	✓			✓	✓	✓		✓	✓	✓			✓	✓	✓		✓		✓	✓	✓	✓	✓		✓		✓
MODLER (Regression analysis)[1]		640K	✓	✓			✓	✓	✓		✓	✓	✓			✓	✓	✓		✓		✓	✓	✓	✓	✓		✓		✓
ABstat-Release 4	Anderson-Bell	196K		✓		✓	✓	✓		✓		✓	✓			✓	✓	✓		✓		✓	✓	✓	✓		✓	✓	✓	✓
DCS MultiStat	Davell Custom Software	192K					✓	✓	✓			✓	✓	✓		✓	✓	✓		✓		✓	✓	✓	✓	✓		✓	✓	✓
Multi	Decision Science Software	32K				✓																							
ESP Model Simulation with 2.0[1]	Economica	512K		✓		✓	✓	✓	✓		✓	✓	✓	✓	✓	✓	✓	✓	✓	✓	✓	✓	✓	✓	✓	✓		✓	✓	✓
ESP 1.2		256K		✓		✓	✓	✓	✓		✓	✓	✓	✓	✓	✓	✓	✓	✓	✓	✓	✓	✓	✓	✓	✓		✓	✓	✓
Exec*U*Stat 1.2	Exec*U*Stat	312K		✓		✓	✓	✓	✓	✓	✓	✓	✓	✓		✓	✓	✓	✓	✓	✓	✓	✓	✓	✓	✓		✓	✓	✓
PC Statistican	Human Systems Dynamics	128K		✓		✓	✓	✓	✓	✓		✓	✓	✓		✓	✓	✓	✓	✓	✓	✓	✓	✓	✓	✓		✓	✓	✓
PC Anova		128K		✓		✓	✓	✓	✓			✓	✓			✓	✓	✓	✓	✓	✓	✓	✓	✓	✓	✓		✓	✓	✓
Multiple Factor Analysis	Mathematical Software Co	64K		✓		✓	✓	✓	✓						✓		✓													
Number Cruncher Stat System	NCSS	196K		✓		✓	✓	✓		✓		✓	✓	✓	✓	✓	✓	✓	✓	✓	✓	✓	✓	✓	✓		✓	✓	✓	✓
NWA Statpak 3.2	Northwest Analytical Inc	128K		✓		✓	✓	✓	✓	✓		✓	✓	✓	✓	✓	✓	✓	✓	✓	✓	✓	✓	✓	✓		✓	✓	✓
Otis	Odin Research	320K		✓		✓		✓	✓		✓	✓	✓	✓		✓						✓				✓		✓	✓	✓
1,2,3 Forecast	1,2,3 Forecast	256K				✓					✓										✓	✓			✓			✓	✓	✓
STATPRO	Penton Software Inc.	192K		✓		✓	✓	✓	✓	✓	✓	✓	✓	✓		✓	✓	✓	✓	✓	✓	✓	✓	✓	✓	✓	✓	✓	✓	✓

(continued)

150

Table 6-3a. (continued).

Product	Vendor	Minimum memory required (in K bytes)	Hard disk required	Integrated Programs	Batch	Interactive	Both	ASCII	DIF	dBASE II	Lotus worksheet	Frequencies	Percentages	Cross tabulation	Factor analysis	Analysis of variance	Multivariate analysis	Linear	Non-linear	Multiple	Time series analysis	High resolution	Histograms	Scatter plots	Line graphs	Bar graphs	Report writing	Data management	Sort files	Join files
Investment Oriented Stat. Software	Programmed Press	64K				✓										✓	✓	✓		✓	✓									
Statistical Stat Pack for Forecasting		128K		✓		✓	✓	✓	✓							✓	✓			✓	✓		✓	✓	✓		✓	✓		✓
The Statistician	Quant Systems	128K		✓		✓	✓	✓	✓				✓	✓			✓	✓		✓	✓		✓	✓	✓		✓	✓	✓	✓
MicroTsp version 4.1	Quantitative Micro Software/McGraw-Hill	256K		✓			✓	✓	✓			✓	✓	✓		✓	✓	✓		✓			✓				✓	✓		
SL MICRO	Questionnaire Service Co	128K		✓	✓			✓				✓	✓	✓				✓	✓	✓	✓	✓						✓		
SmartForecasts	Smart Software Inc	256K	✓	✓					✓	✓			✓	✓	✓		✓	✓	✓		✓		✓	✓	✓	✓	✓	✓	✓	✓	✓
SPSS PC+ Advanced Statistics	SPSS Inc	448K	✓	✓		✓	✓	✓				✓	✓	✓	✓	✓	✓	✓		✓	✓	✓	✓	✓			✓	✓	✓	✓
SPSS PC+ Tables		448K	✓	✓		✓	✓	✓				✓	✓	✓	✓	✓	✓			✓	✓		✓	✓		✓	✓	✓	✓	✓
SPSS PC+		384K	✓	✓		✓	✓	✓			✓	✓	✓	✓	✓	✓	✓	✓	✓	✓	✓	✓	✓	✓	✓	✓	✓	✓	✓	✓
STATGRAPHICS	STSC	384K		✓		✓					✓				✓	✓	✓	✓	✓	✓	✓	✓	✓	✓	✓	✓	✓	✓	✓	✓
Stat 1	Sugar Mill Software Corp	256K		✓		✓	✓	✓	✓			✓	✓	✓	✓	✓	✓	✓	✓	✓	✓	✓	✓	✓	✓	✓	✓	✓	✓	✓
SYSTAT	SYSTAT Inc.	256K		✓		✓	✓	✓	✓			✓	✓	✓	✓	✓	✓	✓		✓			✓	✓			✓	✓	✓	✓
Automatic Statistics	Transaction Systems Inc	128K		✓		✓	✓	✓	✓			✓	✓	✓	✓	✓	✓			✓			✓	✓			✓	✓	✓	✓
Statpac	Walonick Associates	128K		✓		✓	✓	✓	✓			✓	✓	✓			✓	✓		✓				✓			✓	✓	✓	✓
Goodness-of-fit (regression package)		128K		✓		✓	✓	✓	✓									✓							✓					✓
Forecast Plus (Time series analysis)						✓		✓	✓			✓				✓	✓	✓			✓						✓	✓	✓	✓
Statcalc	Zephyr Services	128K		✓		✓		✓	✓	✓	✓	✓							✓			✓							✓	✓	

[1] Requires 8087 coprocessor

151

Table 6-3b. Survey of Statistical Software Compatibility with Different Machines.

	COMPATIBLE MACHINES			
SOFTWARE	IBM PC & 100% COMPATIBLES	WANG	AT&T 6300	APPLES
MOLDER BLUE	YES	YES	YES	NO
ABSTAT-RELEASE 4	YES	YES	YES	NO
DCS MULTI-STAT	YES	NO	YES	NO
ESP	YES	NO	NO	NO
EXEC*U*STAT	YES	NO	YES	NO
HUMAN SYSTEMS DAYNAMICS	YES	NO	YES	NO
MULTIPLE FACTOR ANALYSIS	YES	?	YES	NO
NCSS	YES	?	YES	YES
NWA STATPAK	YES	?	YES	
ODIN	YES	?	YES	NO
1,2,3 FORECAST	YES	YES	YES	?
PENTON	YES			?
PROGRAMMED PRESS	YES	YES	YES	YES
MICRO TSP	YES	NO	YES	YES
QUESTIONAIRE SERVICE	YES	YES	YES	YES
SMART SOFTWARE	YES	YES	YES	YES
SPSS INC				
STSC INC	YES	YES	YES	NO
SUGAR MILL SOFTWARE	YES	YES	YES	NO

compatible programs. This raises two other problems, however. First, when a second program designed for statistics is purchased, the concern of adequate memory in the PC can be serious. With a spreadsheet and companion statistical program, small- to medium-size machines (on the order of 260 K+ memory) can quickly run into memory storage problems. If possible, one can expand the machine's memory, but of course at an expense. The second problem is related to the program language. The majority of statistical software for PC's on today's market is IBM compatible. If, however, one is using another make computer, such as a Wang Corp. PC, one is left out in the cold. In some cases, it may be possible to use a special adapter board to make the PC compatible with other standard software language, but expense and memory capacity return to haunt us. Table 6-3 provides a partial list of software packages with statistics capabilities that were identified at the time this manual was compiled.

Fortunately, many spreadsheet softwares have subroutine commands that enable data to be manipulated in a format suitable for performing regressions. Although a little bit of effort is required to initially adapt a spreadsheet to perform regressions, once the calculations are performed for a sample case, the spreadsheet becomes a template ready for future use. In this section we shall demonstrate how such a template can be constructed on the Lotus 1-2-3. The reader can use an analogous approach to adapt the particular spreadsheet software to his or her needs. As an example, we will use some extrusion data generated from a single-screw extruder for a standard rubber compound.

- Start with a clean worksheet. Use the Worksheet **Erase Yes** command sequence (**/WEY**).
- Type the title **RPM** (for screw RPM) in the A1 cell and set the column width to 5 characters by entering the command **/WCS5** (refer to Table 6-4).
- In a similar fashion, enter **VOL** (for Volumetric Rate) in the B1 column, again using **/WCS5** to establish the column width.
- Allow a space between data titles and the tabulated data. Hence, beginning with cells A3 and B3, type in the data. When typing large bodies of data it is advantageous to repeatedly protect what has been typed by saving it to disk before completing the task.
- Next, move the cursor to cell G3 and type **=n**. Type in the series of expression in cells G4 – G20 as shown in Table 6-4. These are comment statements which identify calculations that are to be performed in cells F3 through F20. As you enter information from Table 6-4, you will notice that the calculations require sums of X^2, Y^2 and the product XY. Use columns M, N, and O to perform these computations. In cell M1 enter the identifying label **X↑2**; in cell N1 type **Y↑2**, in cell O1 type **X*Y**.

Table 6-4. Example of Linear Regression Worksheet.

RPM	VOL	pred Y	2.5	97.5			X^2	Y^2	X*Y	1
										2
0.7	86	92.139	85.714	98.563	45	= n	0.49	7396	60.2	3
1.1	93	94.112	87.753	100.47	178.37	= SUM OF X	1.21	8649	102.3	4
1.05	95	93.866	87.499	100.23	4871	= SUM OF Y	1.1025	9025	99.75	5
1.1	96.5	94.112	87.753	100.47	859.13	= SUM OF X^2	1.21	9312.2	106.15	6
1.9	96	98.060	91.807	104.31	531581	= SUM OF Y^2	3.61	9216	182.4	7
2	95	98.554	92.311	104.79	20058.	= SUM OF X*Y	4	9025	190	8
1.9	100	98.060	91.807	104.31	3.9637	= X MEAN	3.61	10000	190	9
1.6	103.5	96.580	90.291	102.86	108.24	= Y MEAN	2.56	10712.	165.6	10
2.5	98	101.02	94.826	107.21	152.12	= SUM X^2 - ((SUM OF X)^2)/N	6.25	9604	245	11
2.5	100	101.02	94.826	107.21	4322.8	= SUM Y^2 - ((SUM OF Y)^2)/N	6.25	10000	250	12
2.7	100	102.00	95.828	108.18	750.64	= SUM X*Y - (SUM OF X)*(SUM OF Y)/N	7.29	10000	270	13
2.9	100	102.99	96.827	109.16			8.41	10000	290	14
2.7	105	102.00	95.828	108.18	4.9345	= SLOPE b0	7.29	11025	283.5	15
2.2	106	99.540	93.319	105.76	88.684	= INTERCEPT b1	4.84	11236	233.2	16
3.1	105	103.98	97.825	110.13	0.9256	= r* (COEFFICIENT OF CORRELATION)	9.61	11025	325.5	17
3.2	105	104.47	98.323	110.62			10.24	11025	336	18
3.7	105	106.94	100.80	113.08			13.69	11025	388.5	19
3.9	107	107.92	101.79	114.06	3.7931	= ST. DEV. OF POINTS ABOUT REGRESSION	15.21	11449	417.3	20
4.1	107	108.91	102.78	115.05			16.81	11449	438.7	21
4.1	108	108.91	102.78	115.05			16.81	11664	442.8	22
3.99	111	108.37	102.23	114.50			15.920	12321	442.89	23
3.1	111	103.98	97.825	110.13			9.61	12321	344.1	24
3.8	110	107.43	101.29	113.57			14.44	12100	418	25
3.75	115	107.18	101.05	113.32			14.062	13225	431.25	26
4.7	110	111.87	105.72	118.02			22.09	12100	517	27
4.8	110	112.37	106.21	118.52			23.04	12100	528	28
4.8	111	112.37	106.21	118.52			23.04	12321	532.8	29
4.3	115	109.90	103.76	116.04			18.49	13225	494.5	30
4.4	116	110.39	104.25	116.53			19.36	13456	510.4	31
5	117	113.35	107.19	119.52			25	13689	585	32
5.1	112	113.85	107.67	120.02			26.01	12544	571.2	33
5.1	111	113.85	107.67	120.02			26.01	12321	566.1	34
5.2	115	114.34	108.16	120.52			27.04	13225	598	35
5.5	120	115.82	109.62	122.02			30.25	14400	660	36
5.9	115	117.79	111.55	124.03			34.81	13225	678.5	37
6.1	115	118.78	112.52	125.04			37.21	13225	701.5	38
6.05	116	118.53	112.28	124.79			36.602	13456	701.8	39
6.1	118	118.78	112.52	125.04			37.21	13924	719.8	40
6.07	119	118.63	112.37	124.89			36.844	14161	722.33	41
6.5	120	120.75	114.44	127.07			42.25	14400	780	42
6.9	123	122.73	116.36	129.10			47.61	15129	848.7	43
6.85	124	122.48	116.12	128.84			46.922	15376	849.4	44
7.21	124	124.26	117.84	130.68			51.984	15376	894.04	45
7.2	120	124.21	117.79	130.63			51.84	14400	864	46
1	82	93.619	87.244	99.994			1	6724	82	47

A   B   C   D   E   F   G   H   I   J   K   L   M   N   O

- Type the follow calculation commands in the appropriate cells:

Cell:	M3	N3	O3
	+A3↑2	+B3↑2	+A3*B3

Use the Copy command (/C) to fill all the calculations for each column for as many data pairs (rows) typed in earlier.

At this point, those readers more familiar with Lotus 1-2-3 may elect to use the Range Name Create (/RNC) command instead of the Copy command.

- Before performing the regression calculations, construct a scatter plot using the Graph Command. Select the XY option after typing /GT (i.e., the Graph Type command). Choose the x-axis for the first variable with the /GX command. /GB command places the second variable on the y-axis. Command /GOF prevents 1-2-3 from tracing lines between data points. The program will display the menu Graph A B C D E F. Select one of these pending the plotting symbol of preference, and strike the Enter key. A second level menu appears offering Line, Symbol, Both or Neither. Select Symbol and strike Enter key. Striking /GV will enable us to view the scatter plot of column B onto column A.

Title the plot to identify it for future references (e.g., Regression Plot) with the /GOTF command sequence. Using /GOTS one can add a second title (enter /F3 to denote the number of data sets in the analysis). Titles for the X- and Y-axis can also be introduced. Titles will automatically be updated along with the plots when the worksheet is used for computing regressions.

- Returning to the calculation rows in column F (Table 6-4), the following should be entered using the Lotus calculating function symbol command @:

Cell	
F3	@ COUNT (A3 .. AN)
F4	@ SUM (A3 .. AN)
F5	@ SUM (B3 .. BN)
F6	@ SUM (M3 .. MN)
F7	@ SUM (N3 .. NN)
F8	@ SUM (O3 .. ON)
F9	@ AVG (A3 .. AN)
F10	@ AVG (B3 .. BN)
F11	+F6−F4↑2/F3
F12	+F7−F5↑2/F3

Cell	
F13	+F8−F4*F5/F3
F15	+F13/F11
F16	+F10−F15*F9
F17	+F13/@ SQRT (F11*F12)

Note that the end cell in each computation depends on the number of data sets (e.g., if 198 data points are used, then AN = 200).

- The above sequence of calculations produces the coefficients for the regression giving us the linear expression: y = F16 + F15*X. Use this expression to predict any y-value (e.g., volumetric throughput through extruder) for each X (screw speed). To compare predictions with measured values:
  −Enter Predicted Y as a label in cell C1. In cell 3 enter
  +$F$16+$F$15*A3.
  −Using the Copy command, repeat the calculation throughout the column.
  −Update the graphics with the /GA command by typing the range C3 .. CN.
  To distinguish the predicted values from measured data, use the /GOFA command and move the cursor to the Line option in the menu and strike the Enter key. In this manner the predicted y-values will appear as a continuous line and actual data as symbols scattered about the line fit.

- The correlation coefficient is contained in cell F17. This provides a measure of the strength of the correlation.

- Confidence limits control the certainty with which the observed value will fall within a given range of the predicted mean (generated in column C). The size of the range is expressed in standard deviations. Compute the standard deviations of the points around the regression line as follows. Starting with cell F20:

F20 is @SQRT (C1/(F3-2))* (F12-(F13↑2/F11)).

The size of this range determines the probability that an observed value will fall within it. For the extrusion data in Table 6-4, a range of ±3.79 standard deviations encompasses all of the observed values. If this is acceptable, use the above standard deviation in the next calculation. If a stricter confidence limit is desired, e.g., only 5% of correct observations to lie outside the range, the size of the range would have to be decreased to ±1.6 standard deviations.

With cell F20 defined, go to cell D1 and type label 2.5 to denote the 2.5th percentile. In cell D3 type:

+C3 − 1.6*(@SQRT($F$2012*(1+1/$F$3 + (A3-$F$9)↑2/$F$10)))

The dollar sign ($) indicates the absolute reference. Copy D3 into D3 . . DN to return to the lower percentile.

- Repeat the last step and define cell E1 as 97.5. This time type:

+C3+1.6 *(@SQRT($F$2012 *(1+1/$F$3+ (A3-$F$9)↑2/$F$10))).

Copy the contents of E3 into E3 . . EN.

- Update the graph with columns D and E. The final product is a scatter plot showing the best line fit along with the limits for the 2.5 and 97.5 percentile, as shown in Figure 6-9.

Upon completion, we essentially have created a template that can be reused with any data. That is, by retyping new values for X and Y in columns A and B, the regression is automatically performed for us along with a plot.

A similar program can be prepared for semi-logarithmic and exponential regressions. In fact, once the above template is prepared and stored to disk, we can modify the appropriate equations to provide the desired regression formula format, and store the new worksheet under a separate file name. The reader can follow Illustration 3 as a model for establishing a program for a power law fit and Illustration 4 in generating a program for an exponential fit.

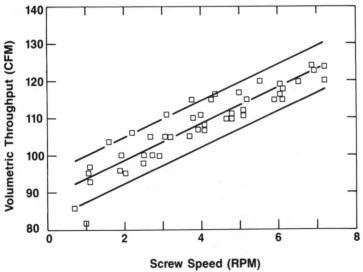

FIGURE 6-9. Plot of extrusion rate in an extruder.

PRACTICE PROBLEMS

1 The following measured values of pressure P of a given mass of gas correspond to various values of the volume V. From thermodynamic principles, the equation of state should have the form $PV^\gamma = C$, where γ and C are constants.

(a) Determine the constants γ and C.

(b) Estimate the pressures for $V = 150, 68, 40$, and 185 in^3 using a regression formula.

Volume, V (in^3)	54.2	61.9	72.5	89.3	119.3	201
Pressure, P(lb/in^2)	61.1	49.4	37.8	29.5	19.9	11.2

2 The U.S. production of paper in millions of pounds during the years of 1976–1983 is given below.

(a) Prepare a time plot of the data.

(b) Prepare an equation of a least square line fitting the data.

(c) Estimate the production during the year 1984.

Year	1976	1977	1978	1979	1980	1981	1982	1983
MM LBS. Paper	100.3	95.4	87.3	89.5	85.5	69.8	65.4	60.5

3 Find the equation of a line passing through the point (7,3) which is parallel to the line $5X + 17Y = 16$.

4 Prove that a least square line always passes through the point $(\bar{X}, \bar{Y})$.

5 The following shows the production of rubber from our four plants in millions of lb. Fit a quadratic expression to these data. (Hint: redefine X so that ΣX and ΣX^3 are zero in order to simplify calculations).

Year	1966	1968	1970	1972	1974	1976	1978	1980	1982	1984
Production (million lb)	20.3	29.4	39.6	53.2	60.5	73.2	97.7	103.3	121.9	135.7

6 A chemical reaction under investigation in a batch reactor appears to be first order. The specific-reaction rate constant is reported below as a function of temperature. (A) Correlate the data in the form of an Arrhenius-type expression (i.e., exponential). (B) Evaluate the activation energy of reaction.

Temp. °C)	295	300	310	323	330	333	339	340
Rate Const., k(min.$^{-1}$)	43.1	42.5	41.2	39.5	38.6	38.4	37.5	37.5

7 Test rats were exposed to different levels of a suspected carcinogen over varying exposure times. Develop a regression equation that correlates the percentage of the population contracting cancer with exposure time and toxicity level per unit body weight. Under what conditions will 50% of the population contract cancer?

Toxicity Level (ppm/kg)	Exposure Time	% Population With Cancer
20	Continuous for 165 days	16
20	8 hrs/day for 3 wks.	13
25	8 hrs/day for 10 wks.	18
35	4 hrs/day for 10 wks.	22
50	4 hrs/day for 10 wks.	63
87	4 hrs/day for 8 wks.	73
90	Continuous for 190 days	77
115	20 min/day for 165 days	69
15	Continuous for 165 days	5
5	Continuous for 190 days	0
7	Continuous for 1 year	2
9	Continuous for 1 year	1

CHAPTER 7

Experimental Design and Parameter Estimation

FACTORIAL DESIGNS

An experimental run is defined by the boundaries of a single combination of factor levels. The experimental design is determined by a set of factor level combinations or experimental runs. A properly designed program should have as many distinct runs as there are model parameters to be estimated. A useful format for structuring an experimental program is the *factorial design*, which is a set of experiments comprising all combinations of levels of each factor.

The simplest class of factorial designs is the *two-level factorial design*. This is usually referred to as the 2^n factorial design in statistics textbooks, and is appropriately named because *n*-factors are considered. In this design, the number of experimental combinations is 2^n.

The value of a factorial design is that it assists in evaluating the effects of several variables simultaneously. It can provide a good prediction of the response over the range of experiments. In discussions to follow, we shall adopt the notation $X_1, X_2, X_3 \ldots$ to denote the design variables. Further, the response of a variable can be coded as positive ($+$) and negative ($-$) to signify higher and lower levels, respectively. A 2^n factorial design will enable us to estimate all the main effects and interactions. To estimate curvature in a response, additional runs are required, and separate discussion is given later. Table 7-1 illustrates a 2^n design for up to five factors. The coded design table is in fact $n \times N$ matrix (n = number of factors, N = number of runs).

The first concern in designing the experiment is the need for replicates. Replicates are repeat runs with identical factor levels. Their functions are to increase the precision of parameter estimates and to provide an estimate of the standard deviation of the measurement. This can be a delicate concern. Too few runs may not be representative of effects. In fact, there's a good chance that experimental error can mask an effect when the population sample is too small. On the other hand, too many experiments not only are time-consuming and wasteful, but may reflect extraneous trends or effects that are of no concern to the study. The following expression enables an estimate of the number of runs needed in a 2^n design (called Wheeler's formula):

$$N \cong \left(\frac{7\sigma}{\Delta} \right)^2 \tag{1}$$

Table 7-1. Illustration of 2^n Factorial Design Layout.

	Run	X_1	X_2	X_3	X_4	X_5
	1	+	−	−	−	−
	2	−	−	−	−	−
2^2	3	−	+	−	−	−
	4	−	−	+	−	−
	5	+	−	+	−	−
	6	+	+	−	−	−
2^3	7	−	−	−	−	−
	8	+	+	−	+	−
	9	−	+	−	+	−
	10	+	−	−	+	−
	11	−	−	−	+	−
	12	+	+	+	+	−
	13	−	−	+	+	−
2^4	14	+	−	+	+	−
	15	−	+	−	−	+
	16	+	−	−	−	+
	17	−	−	−	−	+
	18	−	+	−	+	+
	19	+	+	−	+	+
	20	−	−	−	+	+
	21	+	−	+	+	+
	22	−	+	+	+	+
	23	+	+	−	+	+
2^5	24	−	+	−	+	+

where N = number of experiments in the design, σ = standard deviation (estimated), and Δ = value of the minimum effect we wish to detect with a high degree of confidence (e.g., 5 times out of 6).

As noted in earlier discussions, a well-designed experiment should randomize the run order. This tends to minimize systematic error of time-dependent trends in the process under investigation. In some cases it may be practical to randomize the order in which measurements are taken.

Illustration 1

A new type of fluidized mixer is being considered for a roasting operation. The unit must be able to contact two different density particulates: a fresh solids feed stream and a recycle stream which has already undergone roasting (i.e., a light char). To achieve high throughput capacity we would like to recycle as little of the hot spent stream as possible. The perfect mixer should provide instantaneous mixing of the two streams to complete homogeneity and provide just enough residence time to bring the two streams to equilibrium temperature.

We intend to study the variables affecting the solids residence times in a cold model. The residence time can be related to the following factors:

$$\tau = f(n,\ \nu_g,\ \varrho_{r,f},\ G_{r,f},\ u_g,\ L_s,\ D_s,\ p)$$

where τ = solids mean residence time
    ν_g = kinematic viscosity of the offgas
    $\sigma_{r,f}$ = densities of the recycle and fresh feed solids
    $G_{r,f}$ = mass feed rates of the recycle and fresh feeds
    u_g = superficial velocity of the fluidizing gas through the bed of solids in the mixer
    L_s = length of the screw
    D_s = screw diameter
    p = screw pitch
    n = rotational speed of the mixer screw

There are essentially 10 factors which could affect τ, which means 2^{10} or 1024 possible combinations to test. If we include replicates and assume $\Delta = 3\sigma$, then $N \cong (7\sigma/3\sigma)^2 = 5.4 \sim 6$. We'd be forced to run over 6000 experiments to test all combinations.

A problem like this is best handled by identifying dimensionless numbers which group several of the variables. That is, similarity theory should be applied to reduce the number of variables to a more practical program. The principles of this subject are treated in the text by Cheremisinoff and Azbel (1983) and the derivation for this example problem is given by Cheremisinoff and Cheremisinoff (1984). We shall not review the details other than to state that by application of the Buckingham Pi theorem, the following dimensionless groups can be identified.

$$\theta = f\left(Re_s,\ Fr_m,\ \frac{\varrho_s}{\Delta\varrho},\ D_s/L_s,\ \varkappa,\ \frac{Q_g}{Q_s}\right)$$

where θ is the space velocity defined as $\theta = \tau Q_s/V$
where Q_s = volumetric feed rate of solids
    V = volume of mixer

And the other dimensionless groups are:

$$Re_s = \frac{D_s^2\, n\varrho_{bulk}}{\mu},\ \text{solids Reynolds number with } \mu \text{ gas viscosity.}$$

Q_g/Q_s = gas to solids volumetric throughput.

$Fr_m = \dfrac{\varrho_g p n^2 D_s}{\Delta \varrho\, g}$, modified Froude number.

$\varkappa' = \dfrac{G'}{\varrho_b \nu_g}$, dimensionless solids loading parameter, where G' is the total solids mass flow per unit width of mixer.

The experimental program can now be designed such that we vary the dimensionless groups. Hence, the number of variables required to test all possible combinations has been reduced from 1024 (i.e., 2^{10}) to 64 (or 2^6). In fact, the number of combinations required to test are fewer because some of the dimensionless groups are related to each other, and some variables simply will not vary over significant ranges. For example, offgas viscosity and density are not likely to vary significantly. Hence, if we vary mixer speed, we essentially vary both the solids Reynolds number and Froude number and determine the interaction between these groups simultaneously with a single group of experiments.

FACTORIAL POINTS AND ANALYSIS OF INTERACTIONS

An experimental program should have an exploratory stage aimed at two objectives: first, to define factors or variables affecting a response, and second to

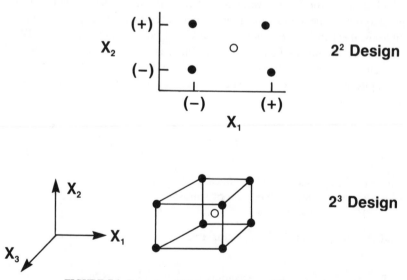

FIGURE 7-1. Examples of 2^n factorial designs with centerpoints.

estimate the overall curvature in the system response. For the latter, this can be accomplished in a 2^n design by adding one or more runs at the centroid of the design. These are referred to by the statistician as the *centerpoint*. The centerpoint response can be compared with the average response of the 2^n design points, which are referred to as the *factorial points*. Figure 7-1 gives examples of 2^n factorial designs with centerpoints. Replication of centerpoints provides an estimate of experimental error. In fact, many times it is preferable to replicate centerpoints rather than to replicate the factorial points, since the latter can distort the balance of the design.

The factorial point responses provide an estimate of the principle effects as well as the interactions. Such an analysis can be readily carried out as a spreadsheet computation. The following outlines the general approach along with an illustration. The general idea is to take each principle effect factor and average it over all the other factors. By utilizing all of the data in estimating each effect, we obtain maximum precision for each estimate.

- Tabulate the design, including the appropriate run numbers for respective X_i columns. The table should include the response values (or response averages if measurements were replicated) for each run.
- Compute a column for $X_i X_j$ interaction.
- As in the ANOVA, sum the values in the response column and enter this in the positive summation row ($\Sigma +$). Divide this by the number of runs to obtain an overall average and record this in the "EFFECTS" row.
- Perform the following computations for each main effect and interaction:
  - sum the corresponding "+" response and record in the "$\Sigma +$" row.
  - sum the "−" corresponding responses and record in the "$\Sigma -$" row.
  - compute the difference between $\Sigma +$ and $\Sigma -$ values.
  - divide the difference by one-half the number of runs. This value belongs in the "EFFECT" row, as it represents an estimate of the main effect or interaction.

Illustration 2

We shall continue with the system described in Illustration 1. From mixing practices/theory, we already know that screw speed and L/D_s (i.e., Froude number and Reynolds number) will have a dramatic effect on residence time. Therefore, our initial efforts will be aimed at scoping out the effects of solids loading and fluidizing gas velocity on mixer performance.

Table 7-2 reports data on the effect of dimensionless solids loading (i.e., effect of solids feed) and the gas to solids throughput ratio on τ for one size mixer. Hence, all other variables described in Illustration 1 are kept constant and we have designed a 2^2 factorial, and the variables to evaluate are X_3 and

Table 7-2. Data for Illustration 2.

Run No.	Solids Feed (kg/hr)	Mixer Speed (RPM)	(X_1) Solids Reynolds No.	(X_2) Solids Froude No.	(X_3) Dimensionless Loading, k	(X_4) Gas to Solids Throughput	(Response) Residence Time (sec.)
1	936	124	993,797	0.21	105.8	28.5	11.9
2	2150	124	993,797	0.21	243.0	15.0	11.3
3	2150	124	993,797	0.21	243.0	22.9	11.0
4	1675	124	993,797	0.21	189.3	17.2	12.0
5	860	124	993,797	0.21	97.2	8.0	18.1
6	1150	124	993,797	0.21	130.0	6.0	20.5

X_4. Following the steps outlined above, we obtain:

Run No.	X_3	X_4	Response	Main Effects X_3	Main Effects X_4	Interaction X_3X_4
1	105.8	28.5	11.9	−	+	−
2	243.0	15.0	11.3	+	−	−
3	243.0	22.9	11.0	+	−	−
4	189.3	17.2	12.0	+	−	−
5	97.2	8.0	18.1	−	−	+
6	130.0	6.0	20.5	+	−	−
		$\Sigma+$	84.8	54.8	11.9	18.1
		$\Sigma-$		30.0	72.9	66.7
		Δ		24.8	−61	−48.6
		Effect	14.13	8.27	−20.3	−16.2

The "EFFECT" row tells us the following: The overall average residence time is 14.1 sec. The effect of raising the solids loading (or solids feed) over the range tested was to increase the residence time by 8.3%. The average effect of lowering the gas throughput (or gas superficial velocity) was to decrease the residence time by 20%. The interaction of gas throughput and solids feed is negative; hence, the residence time−solid's loading slope decreases with decreasing gas throughput (see Figure 7-2 for expected relationship). From a hydrodynamic standpoint, the data reveal that as loading increases, the bed height in the mixer increases, and so at constant screw speed we would expect the solids to take longer to travel through the unit.

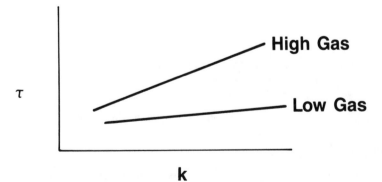

FIGURE 7-2. Expected interactions for Illustration 2.

To estimate the confidence intervals the method outlined below can be used. The procedure assumes that an estimate of the experimental error is available through replication of runs. If the factorial design is replicated then each factorial point provides an estimate of the error. These estimates can be checked for homogeneity and pooled to provide an overall estimate. The familiar Student's t statistic can be used for calculating the confidence interval on each effect, based on ν degrees of freedom. The standard error of the effect can be calculated from:

$$SE = S_p/\sqrt{N/4} \qquad (2)$$

where S_p = pooled estimate of error with ν degrees of freedom. The analysis procedure is as follows:

- Compute the main effects in interactions as illustrated in the procedure outlined for Illustration 2.
- Compute the pooled estimate of error S_p using the procedure given in Chapter 4. For duplicates, use:

$$S^2 = \Delta^2/2 \qquad (3)$$

where Δ = difference between duplicates.

- Obtain t from Table 2-2 in Chapter 2 for the desired confidence level and degrees of freedom on S_p. For r replicates of a 2^n design, the degrees of freedom are:

$$\nu = (r - 1)2^n \qquad (4)$$

- Compute the confidence interval as follows:

$$\text{Effect} \pm tS_p/\sqrt{N/4} \qquad (5)$$

where $N = r2^n$ (the number of experiments)
r = number of replicates

The procedure is demonstrated in the following illustration.

Illustration 3

Fly ash is being considered as a potentially inexpensive replacement for activated carbon in adsorption for water pollution control practices. Batch laboratory tests are being conducted to assess the adsorption efficiency of fly ash. Using the same initial concentration of compund A in water, a 2^2 factorial experiment in duplicate is run to estimate the effects of contact time and temperature on removal efficiency of compound A. The data are:

Run	Temp. (°C)	Time (hrs.)	% Removal (mass basis)
1	28	0.5	48,59
2	50	0.5	73,79
3	28	1.5	80,80
4	50	1.5	89,84

Let's evaluate the main effects and determine the confidence intervals on the effects.

The estimation of the main effects is as follows:

Run	X_1	X_2	Response	Main Effects X_1	X_2	Interaction X_1X_2
1	28	0.5	53.5	−	−	+
2	50	0.5	76.0	+	−	−
3	28	1.5	80.0	−	+	−
4	50	1.5	86.5	+	+	+
		$\Sigma+$	296.0	162.5	166.5	140.0
		$\Sigma-$		133.5	129.5	156.0
		Δ		29.0	37.0	−16.0
		Effect	74	14.5	18.5	− 8.0

The overall average removal of compound A is 74%. The average effect of raising the temperature from 28 to 50°C is to increase removal efficiency by 14.5%. The average effect of raising the time from 30 min. to 1½ hrs. is to increase removal efficiency by 18.5%. The interaction of time and temperature is negative; hence, the removal efficiency-temperature slope decreases as contact time increases.

We can now estimate the confidence intervals on the effects:

$$X_1 = 14.5, \ X_2 = 18.5 \text{ and } X_1X_2 = -8.0$$

$$S_p^2 = \frac{60.5 + 18 + 0 + 12.5}{4} = 22.75$$

$S_p = 4.77$ with 4 degrees of freedom.

From Table 2-2 (Chapter 2), for 95% confidence $t = 2.776$, with 4 degrees of freedom.

$$N = r2^n = 2(2^2) = 8$$

$$\therefore tS_p/\sqrt{N/4} = (2.776)(4.77)/\sqrt{2} = 9.36$$

Hence,

for temperature:	14.5 ± 9.36 (or 5.14 to 23.86)
for time:	18.5 ± 9.36 (or 9.14 to 27.86)
for interaction:	−8.0 ± 9.36 (or −17.4 to 1.36)

These are the 95% confidence intervals. Since the intervals are broad, we might well consider re-running the tests or taking a closer look at the reproducibility of run 1.

We now direct attention to assessing the overall curvature of the response. For a 2^n factorial design, an estimate of the overall curvature due to one or more factors can be made by running an estimate of the experimental error, which can be pooled with any error estimate from the factorial points. The following procedure can be used:

- Compute the average centerpoint response and the standard deviation of the centerpoint replicates. Pool the standard deviations with the factorial points.
- Apply the Student's t-statistic by using Table 2-2 in Chapter 2 for the specified confidence level and pooled error degrees of freedom.
- An estimate of the response's curvature can be made by computing the difference between the average centerpoint response, X_c, and the average of the factorial points, X_f:

$$\text{Curvature} = \bar{X}_c - \bar{X}_f \tag{6}$$

The confidence interval on the curvature can be estimated from the following limits:

$$\text{Curvature} \pm tS_p\sqrt{1/N' + 1/c'} \tag{7}$$

where
N' = number of factorial points
c' = number of centerpoints

Illustration 4

In the test program of Illustration 3, two centerpoint runs were made (i.e., at 39°C and 1.0 hrs). The adsorption efficiencies measured were 81 and 83%.

Computing the average centerpoint response, we get:

$$\bar{X}_c = (81 + 83)/2 = 82\% \text{ removal.}$$

$$S_c^2 = (2)^2/2 = 2$$

$$\therefore \quad S_c = 1.41 \text{ with 1 degree of freedom.}$$

Pooling with the previous estimate ($S = 4.77$ with 4 degrees of freedom) results in:

$$S_p^2 = \frac{4(4.77)^2 + 1(1.41)^2}{5} = 18.60$$

$$S_p = 4.31 \text{ with 5 degrees of freedom.}$$

From Table 2-2 (Chapter 2) for $\alpha = 0.05$, $t = 2.571$. From Illustration 3, $\bar{X}_f = 74.0$, hence, the curvature is:

$$\text{Curvature} = 82 - 74 = 8\%$$

The 95% confidence interval is:

$$8 \pm (2.571)(4.31) \sqrt{\frac{1}{8} + \frac{1}{2}}$$

$$8 \pm 5.98, \text{ or } 2.02 \text{ to } 13.98.$$

SCREENING TESTS AND OVERALL DESIGN

When a large number of variables that potentially could affect the response exist, the use of screening tests should be considered. The objective of a screening test is to identify the most important factors. It therefore represents the first stage of an experimental program. Part of the exercise involves listing all potential factors along with the extremes of their limits (i.e., their high and low levels). To achieve good sensitivity over the range of experimental conditions, the levels should be as wide as is practical. Remember, the emphasis in this part of the experimental design is to identify main effects. Interactions should not be estimated on an individual basis at this stage; however, group estimates can be informative. With multi-variable systems, often interactions are

Table 7-3. 8-Run Fractional Factorial Designs.

Run	X_1	X_2	X_3	X_4	X_5	X_6	X_7
1	−	−	−	−	+	+	+
2	+	−	−	+	+	−	−
3	−	+	−	+	−	+	−
4	+	+	−	−	−	−	+
5	−	−	+	+	−	−	+
6	+	−	+	−	−	+	−
7	−	+	+	−	+	−	−
8	+	+	+	+	+	+	+
3 Variables	1	2	3	(123)	23	13	12
4 Variables	1	2	3	4	23	13	12
					14	24	34
5 Variables	1	2	3	4	5	13	12
	45	35	25	15	23	24	34
					14		
6 Variables	1	2	3	4	5	6	12
	45	35	25	15	23	13	34
	36	46	16	26	14	24	56
7 Variables	1	2	3	4	5	6	7
	45	35	25	15	23	13	12
	36	46	16	26	14	24	34
	27	17	47	37	67	57	56

masked by or compounded with main effects. In other words, these estimates may be correlated. Mathematically speaking, we may model multi-variable systems with a global expression:

$$Y = b_o + b_1 X_1 + \ldots + b_n X_n + \text{Interactions} + \text{Curvature} \qquad (8)$$

Plackett and Burman (1947) explain the basis for factorial design selection and the use of screening tests. We shall make only a few general comments concerning this subject and refer the reader to this reference. The choice of the design depends on the number of factors n and the total number of runs N required for precision. Furthermore, the design should be chosen such that $2 \sim 5$ factors are left as "dummy" factors. These can be used as a measure of overall interaction or as an estimate of error in the absence of any replication. As an example, Table 7-3 shows an 8-run fractional factorial design and the possible interaction combinations. This type of table illustrates the nature of confounding between interactions and main effects. Table 7-3 represents a simple design case and of course represents only one of many possible design configurations. For many analyses, designs can be based on multiples of four runs, ranging from 8 runs to 24 runs. Larger designs can be found in the literature (see Suggested Readings on page 203).

Once the main variables have been identified and their effects estimated, emphasis should be placed on developing a model that enables prediction of the point of optimum response. Statistics textbooks refer to this subject as *response surface designs*, and the approach involves the use of an n-factor second-order empirical model of the form:

$$Y = \beta_o + \Sigma\beta_i X_i + \Sigma\beta_{ii} X_i^2 + \Sigma\beta_{ij} X_i X_j \qquad (9)$$

Subscripts i and j denote the indices from 1 to n and $i + 1$ to n, respectively.

The experimental design should provide us with enough information to estimate the model parameters. The optimal response can then be assessed either analytically, or graphically. A good design is one which adequately covers the experimental region of interest. In addition, it should be symmetric about a centerpoint. To assess all possible factors, the number of distinct factor level combinations in the design should be at least equal to the number of model parameters. Furthermore, at least three factor levels should be used for each factor in order to estimate the β_{ii} curvature parameters. A final criterion is that the design should be based on a factorial scheme in order to estimate the β_{ij} interaction parameters.

Central composite designs are an extension of the 2^n factorial design or fraction. This type of experimental design greatly increases the precision of the analysis through the use of *star* points. The star points are illustrated by the interaction plot between two factors shown in Figure 7-3. For $2n$ star points, one factor is maintained at $\pm\alpha$, with the other factors held at midpoints. Hence, the design provides us with more than one centerpoint. The number of centerpoints depends on the number of factors. For 2 factors there are 4 centerpoints, for 3 factors—6 centerpoints, 4 factors—7 centerpoints, 5 factors—10 centerponts. For rotability, $\alpha = 2^{n/4}$ for a full factorial design. An example of a central composite design is illustrated in Figure 7-4 for a 3-factor model. In this example, the factorial design is comprised of:

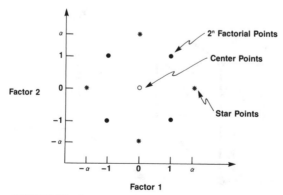

FIGURE 7-3. Illustrates the use of star points in the design.

Symbol Used in Figure 7-4		Factor		
		X_1	X_2	X_3
○	Centerpoints	0	0	0
●	Factorial Points	−	−	−
		+	−	−
		−	+	−
		+	+	−
		−	−	+
		+	−	+
		−	+	+
*	Star Points	0	0	$-\alpha$
		0	0	α
		0	$-\alpha$	0
		0	α	0
		$-\alpha$	0	0
		α	0	0

The reader should refer to the work by Gorman and Hinman (1962) and Narcy and Renaud (1972) to obtain detailed discussions of multi-factor designs.

To estimate the parameters in the full quadratic model, multiple regression analysis must be applied. This involves the use of a computer program that is capable of computing the standard deviation of the parameters, provides a tabulation of the residuals (i.e., the differences between experimental values of the response variable and values predicted by the model), and provides plotting routines that show residuals vs. predictions and residuals vs. X variables. It is useful in such an analysis to code the X-values prior to estimation. Several of the commercial programs described in Table 6-3 of Chapter 6 provide multiple regression capabilities for up to 6 or 7 variables. In most of these programs, graphical presentation of the regression model is possible, which is

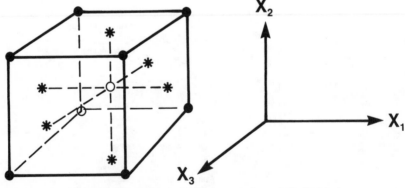

FIGURE 7-4. Illustrates a 3 factor central composite design.

highly useful when reconciling two or more responses. Contour plotting routines provide contours of Y plotted against two X-variables at fixed values of other X variables in the model. If possible, it's always useful to superimpose data values on the contour plots. This will provide a rough idea of the model fit.

The commercial programs listed in Chapter 6 are user friendly, and most authors have gone to great lengths to provide descriptive documentation on their usage. It would be out of context to recommend one system over another, particularly since selection will largely be based on the types of statistical problems the reader most frequently encounters. There are sufficient differences among the varieties of analysis in the commercial packages that the reader may wish to try and categorize the types of problems he or she most often encounters. As a first pass, for multiple regression, one can apply a program such as "DATA-FIT" described in Appendix A by holding one of the variables constant. This routine, designed for LOTUS 1-2-3 users, can evaluate up to three variables.

Simplex experimental designs can be applied to multicomponent systems. A concerns the properties of mixtures of several components. The overall properties in this case depend on the relative percentages of the components as well as their total amount. This is typically the case when several polymers are blended for definite usage conditions. Another common example of this class of problems is the formulation of detergents. In these cases we are interested in the mixture's performance or physical properties/characteristics. A proper experimental design will result in a set of polynomial equations and graphical representations which enable one to predict responses for a range of different mixtures.

A simplified procedure for analyzing this problem class has been proposed by Narcy and Renaud (1972). Because of its simplicity and ease in application it is recommended for use, and discussions illustrating this methodology are given below.

The approach is based on *response surface designs*. If our objective is to develop an empirical model to predict a particular property, then what we seek is an equation that is likely a low order polynomial comprised of only a few terms. The model can be expanded by simply adding more terms (or variables) to improve the fit. Hence, a simplex design expression will have the form:

$$Y = \beta_1 X_1 + \beta_2 X_2 + \beta_3 X_3 + \beta_4 X_1 X_2 + \beta_5 X_1 X_3 + \beta_6 X_2 X_3 + \epsilon \quad (10)$$

where β_i's are the coefficients, X_i's are the component proportions and ϵ is an infinitesimal Gaussian variate.

Ignoring ϵ gives $\hat{Y}$ which is the prediction of the response Y. The resulting expression represents the response surface of the system.

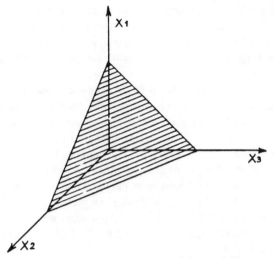

FIGURE 7-5. Simplex representation of three variables.

There is no general graphical representation for three variables; however, when applying the simplex design, we can obtain a specific plot for illustrative purposes. Consider the domain of (X_1, X_2, X_3) components or mixtures, such that $X_1 + X_2 + X_3$ = constant. The definition of the simplex is:

$$X_1 + X_2 + X_3 = 1$$

$$X_1 > 0, X_2 > 0, X_3 > 0$$

(11)

Figure 7-5 shows the representation. The constraint reduces the parameter space dimensionality from three to two. This means that (X_1, X_2, X_3) lie inside the triangle. Four components result in a tetrahedron.

The Scheffe method (1958, 1963) explores the components' space at points corresponding to an ordered arrangement known as a lattice. Two families of designs are available. First, the simplex lattice design is determined by two figures: q for the number of components, and m for the degree of the polynomial response. Design points consist of pure components and mixtures of two or more components (up to m). For each mixture the proportions of the components are simple fractional numbers, design points symmetrically covering the simplex. When there is only one measurement per point (one replication), the sample size is given by the formula: $n = (m + q - 1) !/m!$ $(q - 1)!$ Examples of representations of several (q,m) simplex-lattice designs are given in Figure 7-6.

Scheffe (1963) also introduced the simplex-centroid. Contrary to the lattice design, in which there is no mixture with more than m components, centroid designs are such that for any number of components between 1 and q there exist one or several design points corresponding to these mixtures. In particular there is one point with all components present in equal proportions. The sample size for one replication is $(2q - 1)$ (see Figure 7-7).

The simplex centroid is relevant for mixtures of many components when response is simple (low order). The simplex lattice would apply to higher order variations but with not too many components, because estimation of regression coefficients becomes tedious when q is large.

Both families provide polynomial equations, the coefficients of which are simple functions of the measured responses at the experimental points.

Scheffe (1963), and Gorman and Hinman (1962) show the exact forms of the response surfaces along with the linear equations that produce the polynomial coefficients as functions of the responses at the design points. These studies were done for three and four components and the references should be consulted for details. For illustration, consider the cubic simplex lattice design with three components.

3 COMPONENTS

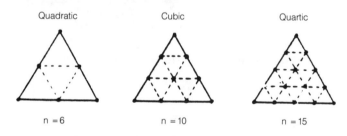

4 COMPONENTS

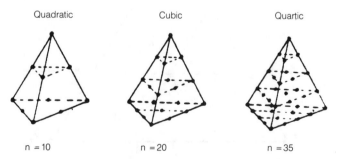

FIGURE 7-6. Simplex lattice designs.

3 COMPONENTS
n = 7

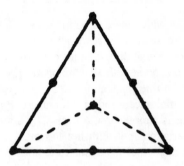

4 COMPONENTS
n = 15

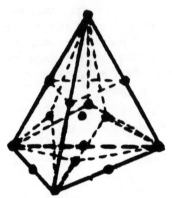

FIGURE 7-7. Simplex centroid designs.

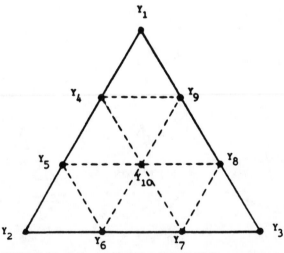

FIGURE 7-8. Cubic simplex lattice design with three components.

There are 10 design points, and 10 coefficients are to be estimated. Let Y_i denote the responses, with i as indicated on Figure 7-8.

The response surface is:

$$Y = \beta_1 x_1 + \beta_2 x_2 + \beta_3 x_3 + \beta_{12} x_1 x_2 + \beta_{13} x_1 x_3 + \beta_{23}$$
$$x_2 x_3 + \gamma_{12} x_1 x_2 (x_1 - x_2) + \gamma_{13} x_1 x_3 (x_1 - x_3) \quad (12)$$
$$+ \gamma_{23} x_2 x_3 (x_2 - x_3) + \beta_{123} x_1 x_2 x_3.$$

β_i's and γ_i's respectively are determined as follows:

$$\beta_1 = Y_1$$
$$\beta_2 = Y_2$$
$$\beta_3 = Y_3$$
$$\beta_{12} = (9/4)(Y_4 + Y_5 - Y_1 - Y_2)$$
$$\beta_{13} = (9/4)(Y_8 + Y_9 - Y_2 - Y_3)$$
$$\beta_{23} = (9/4)(Y_6 + Y_7 - Y_2 - Y_3) \quad (13)$$
$$\gamma_{12} = (9/4)(3Y_4 - 3Y_5 - Y_1 + Y_2)$$
$$\gamma_{13} = (9/4)(3Y_9 - 3Y_8 - Y_1 + Y_3)$$
$$\gamma_{23} = (9/4)(3Y_6 - 3Y_7 - Y_2 + Y_3)$$
$$\beta_{123} = 27Y_{10} - (27/4)(Y_4 + Y_5 + Y_6 + Y_7 + Y_8 + Y_9)$$
$$+ 9/2 (Y_1 + Y_2 + Y_3).$$

The regressed response surface must be checked for its adequacy to the true responses. This is done with tests of fit, which deliver proper measures comparing the bias error to the experimental error.

By construction of this type of design, there is a perfect fit of the model at all design points. Consequently, check points are to be different from these points. Some of them will correspond to mixtures of a particular interest, whereas others will consist of design points of different simplex models, thus permitting economy in experimentation.

The number of check points necessary depends on the degree of the response polynomial. One must perform the fit test for each check point and report the individual results, fit or no fit, on the graphical representation of the design.

By examining the graphical representation one gets an overall mapping of the adequacy of the tested model. The procedure [described by Fisher and Yates (1967)] follows.

Suppose we run r different measurements at the check mixture, $\bar{Y}$ denoting the average measured response. Let $\hat{Y}$ denote the predicted response at that point, as given by the model under test. What follows relies on two basic assumptions: first, at each of the n design points, all observations have been equally replicated, say k times. Scheffe's theory may apply to a more general case where observations can be replicated differently, according to the position of the mixture, vertex, midpoint, centroid, etc. However, for the user's convenience and simplicity in calculation, equal replication is better. The second assumption concerns the experimental error (σ^2). Assume that σ is homogeneous enough to be considered constant over the domain of the study. The test procedure consists in comparing the characteristic $\tau_{f,g,\upsilon}$ with the critical value given in Behrens tables [see Fisher (1967)]:

$$\tau_{f,g,\upsilon} = |\bar{Y} \cdot \hat{Y}| / \sigma\sqrt{1/r + \xi/k} \tag{14}$$

where f and g are the numbers of degrees of freedom, respectively, associated with $\bar{Y}$ and $\hat{Y}$: $f = r - 1$; $g = nk - 1$; and θ: $\tan\theta = \sqrt{\xi(r/k)}$.

ξ is an important parameter which is a symmetric function in the X_i's and only depends on the position of the check point. Contour plots for ξ are given

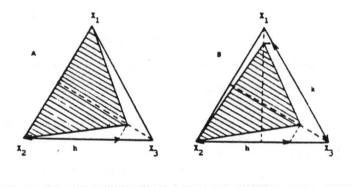

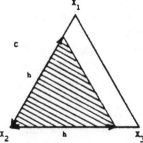

FIGURE 7-9. Examples of restricted designs: (a) one extreme vertex; (b) two extreme vertices; (c) one extreme edge.

by Gorman (1962). We use the Behrens test rather than the usual Student test for paired comparisons, because the errors on $\bar{Y}$ and $\hat{Y}$ arise from different sources. Only the Behrens test properly accounts for this situation, by introducing an additional parameter, $\tan\theta$, defined as the ratio of two variances.

$$\tan\theta = \sqrt{\operatorname{var} \hat{Y}/\operatorname{var} \bar{Y}} \text{ with: } \begin{aligned} \operatorname{var} \hat{Y} &= \xi(\sigma^2/k) \\ \operatorname{var} \bar{Y} &= \sigma^2/r \end{aligned} \tag{15}$$

When $f = g$ and $\theta = 45°$, Behrens and Student procedures are identical. With $f = g$ but $\theta \neq 45°$, the Student test would be too severe. With $f < g$, the Student test would be too severe for $\theta < 45°$ and not severe enough for $\theta > 45°$.

After the fit test has been performed on all check points we have a good idea of the fit of the model. A frequent cause of misfit is the presence of one or several extreme values, very different from the response at neighboring points, on the boundaries of the domain. In this case handling the design remains possible, with appropriate changes in coordinates. These changes act as if the domain were restricted (see Figure 7-9). The shaded areas represent the reduced domains, which are simplexes in X_i'.

Narcy and Renaud (1972) have generalized a technique, first recommended by Scheffe for different purposes, which is called the pseudovertex method. For case A, if the extreme value corresponds to the third component, $X_3 = 1$, we search for a point, with proportion h of this component, such that the response becomes reasonable, by gradually reducing X_3. Several trials may be necessary to find the most appropriate proportion h.

Coordinates of the pseudovertex are:

$$\begin{cases} x_1 = x_2 = (1 - h)/2, x_3 = h \\ 0 \le x_3 \le h < 1 \end{cases} \tag{16}$$

This leads to the following change in coordinates:

$$\begin{cases} \text{(I)} \quad \begin{aligned} x_1 &= x'_1 + [(1 - h)x'_3/2] \\ x_2 &= x'_2 + [(1 - h)x'_3/2] \\ x_3 &= hx'_3 \end{aligned} \\ \\ \text{(II)} \quad \begin{aligned} x'_1 &= x_1 - [(1 - h)x_3/2h] \\ x'_2 &= x_2 - [(1 - h)x_3/2h] \\ x'_3 &= x_3/h \end{aligned} \end{cases} \tag{17}$$

One can check that $X'_1 + X'_2 + X'_3 = X_1 + X_2 + X_3$. Consequently, the reduced design is still a simplex, but in (X'_1, X'_2, X'_3).

System I is first used, in order to determine the true mixtures to be run for the new design.

System II is used for testing the lack of fit and for prediction of responses at mixtures of interest.

Case B, in Figure 7-9, can be solved with two successive changes in coordinates which result in:

$$\begin{cases} x'_1 = (x_1/k) - [(1 - h)x_3/2hk] \\ x'_2 = x_2 - [(1 - k)x_1/2k] - [(1 - h) x_3/2h] + [(1 - k) (1 - h)x_3/4hk] \\ x'_3 = (x_3/h) - [(1 - k)x_1/2k] + [(1 - k)(1 - h)x_3/4hk]. \end{cases} \quad (18)$$

Finally for case C:

$$\begin{cases} x'_1 = x_1/h \\ x'_2 = x_2 - [(1 - h)(x_1 + x_3)/h] \\ x'_3 = x_3/h \end{cases} \quad (19)$$

The latter system is about as simple as for case A, but it may be handled in a different way. As it appears on Figure 7-9c the restricted simplex is an equilateral triangle. With a suitable h we may avoid the change in coordinates. For example, let us consider a quartic design; if $h = 0.75$, the remaining design points correspond to the cubic model for the restricted design.

Besides, it may happen that the domain of interest is very different from an equilateral triangle. But very often this domain can be split into several adjacent equilateral triangles. In particular, an isosceles trapezium such that one base line is twice the other can be split into three equilateral triangles. Actually, this case arises when we cannot formulate the upper half of a Scheffe's simplex, since the other half is an isosceles trapezium.

Depending on the application, the number of components can vary from three to about ten. When the number of components becomes large (e.g., $n > 5$), only a quadratic fit is reasonable. This is the case in the petroleum industry, where problems such as octane-blending vs. performance characteristics are considered. Table 7-4 lists the number of design points for up to 5 components. Note that two cells are crossed out. In one case the number of ex-

Table 7-4. Required Number of Design Points.

q	Quadratic	Centroid	Cubic	Quartic
3	6	7	10	15
4	10	15	20	35
5	15	31	35	70

periments is probably prohibitive; in the other, the crossed design is identical to the next one, with the exception of one point. It should not be thought that increasing the degree is necessarily an improvement.

Narcy and Renaud (1972) note that cubic and quartic deviations from the experimental response are of opposite sign; the quartic model is closer to the experimental situation than the cubic one; the distance between cubic and quartic curves is enlarged by the presence of point 7, which has an extreme value; and by taking off point 7 (restricted simplex) a quadratic model would fit best. In practice the experimenter, facing a lack-of-fit situation with the cubic model, would have tried the quartic one with only a very small benefit. Also, a cubic design should be tried when a minimum and a maximum are expected, whereas a quartic design should be run for three extrema. The steps required for a sequential search of proper simplex designs are these. The first step (a) consists in defining the domain to be considered. The limits of the domain are subject to three constraints: formulation feasibility, mixtures of interest from an economic point of view, and responses in the ranges of interest. A few pilot experiments are necessary. A second step (b) is to measure the responses at the vertices: the pseudovertex feature is used in case of extreme values. A third step (c) is to run a simple model, that is the quadratic lattice or the centroid. Measurements at the design points are replicated to reach a low variance for polynomial coefficients, and we establish the polynomial equation of the model. Check points are chosen lying inside the domain, such as the inner points of the quartic model. If fit is correct, graphical representations are drawn for systems of three components. Optimal areas are isolated. If we have to consider several responses, an "optimum optimorum" is obtained by superposition of the respective graphs.

If the model is inadequate, several decisions can be made: if many discontinuities are encountered, either reduce the domain or split it in several parts; if values appear to be very abnormal, reconsider the problem, gain knowledge of the phenomenon and possibly try an "explicative" model, perhaps nonlinear. Otherwise, move to the cubic lattice design. Proceed similarly for cubic and eventually for quartic. Checkpoints for the cubic model will be chosen roughly half among design points of the previous model, half among those of the quartic model. For the checkpoints of the quartic model, try all cubic points that do not belong to the quartic design, together with some inner points of interest. Narcy and Renaud give examples which illustrate this procedure.

CLOSURE AND GENERAL COMMENTS ON EXPERIMENTAL DESIGN

An experimental design should follow a logical progression from inception to end. The general steps that constitute the program are:

- Exploratory Stage—This initial stage of the design should be aimed at scoping out the objectives (i.e., What do we want to learn from the study? Is the overall objective to develop a scale-up model?). During this period, the experimenter should become familiar with the apparatus (i.e., the equipment, instrumentation, sensors, etc. should be checked out). This period should be used to assess the precision of instruments as well as the accuracy of measurements to be made. Estimates of the experimental error can be obtained during this stage.

- Variable Screening Stage – Preliminary experiments should be conducted to scope out the range of the experimental region. Most important, one should identify major factors affecting the response or responses under investigation. This part of the program has the overall objective of reducing the large number of potential variables to the most effective ones.

- Optimization Scoping Stage – More detailed experiments can now be implemented to better define the experimental region containing the optimum response. Once this region is defined, replicates can be run to obtain estimates of the standard deviation.

- Final Optimization and Modelling Stage – Effort should be placed on developing a predictive model for the response in the optimum region. At the optimum point, the 2^n factorial design can be expanded into a central composite design to estimate the parameters of a full quadratic model. This will provide a basis for interpolation of the response.

To this last point, the final product we obtain is likely an empirical model that should provide a good description and prediction of the response phenomenon over the experimental region. Note, however, that the model must have a theoretical basis if it is to be a useful tool. If the response model provides predictions which are contrary to physical and/or physio-chemical laws, then any degree of correlation obtained can likely be attributed to chance or unidentified factors. One should, therefore, always attempt to explain the resulting correlation in terms of theoretical considerations.

The objective of this monograph was to provide a handy reference of short-cut statistical procedures along with some guidelines for data analysis and interpretation. We have purposely sidestepped the theorems behind these methods and refer the reader to the references cited at the end of the book for more detailed discussions. Appendix B provides definitions of frequently used statistical terms and will acquaint the less experienced reader with some of the terminology found in the more definitive literature.

PRACTICE PROBLEMS

1 Studies were conducted to assess the parameters affecting minimum fluidization velocity (i.e., the superficial gas velocity at which particles first become fluidized). The tests were conducted for

different materials of varying sizes and over a range of pressures. The fludizing medium was N_2 at 28°C. Determine the main variables affecting minimum fluidization conditions from the data below.

Material	Particle Density	Mean Particle Size (μ)	Pressure (kPa)	Minimum Fluidization Velocity, μ_{mf} (cm/s)
Coal	0.92	86	150	0.6
			2100	0.5
			4200	0.6
		181	150	2.0
			2100	1.8
			4200	1.85
		301	150	6.6
			2100	5.9
			4200	4.6
Char	0.87	96	150	0.7
			2100	0.6
			4200	0.65
		196	150	6.6
			2100	5.8
			4200	4.5
Sand	1.67	102	150	1.1
			2100	0.9
			4200	0.8

2 The use of ozone in conjunction with lime has been investigated as a manganese removal technique from water. The data below [data extracted from Patterson, J. W. *Wastewater Treatment Technology*. Ann Arbor Science Pub. Co., Inc., Ann Arbor, MI (1975)] are results for solutions of manganous chloride initially containing 94 mg/l manganese. Assess the validity of ozone treatment and the confidence limits.

pH	Final Manganese Conc. (mg/l)	
	Lime Only	Lime Followed by Ozone
7.0	—	0.8
8.0	90	0.3
9.0	17	≤0.1
10.0	0.15	≤0.1
11.0	≤0.1	—

3 Data on electrolytic decomposition of cyanide waste for wastewater treatment [from Easton, J. K., *J. Water Poll. Cont. Fed.*, *39*, 1621–1625 (1967)] are reported below. Assess the interaction between cyanide concentration and decomposition time. Develop a correlation to predict final cyanide concentration.

Run No.	Initial Cyanide Conc. (mg/l)	Time to Decompose (days)	Final Cyanide Conc. (mg/l)
1	95,000	16	0.1
2	75,000	17	0.2
3	50,000	10	0.4
4	75,000	18	0.2
5	65,000	12	0.2
6	100,000	17	0.3
7	55,000	14	0.4
8	45,000	7	0.1
9	50,000	14	0.1
10	55,000	8	0.2
11	48,000	12	0.4

"DATA-FIT" Program User's Guide

GENERAL DESCRIPTION

DATA-FIT is a user friendly statistical software package which runs on the LOTUS 1-2-3 spreadsheet. It is ideally suited for practitioners who need quick data analysis but do not have the time to involve themselves with the lengthy and repetitive calculations which most regressions require.

DATA-FIT contains five different regression models which users can fit to their data. These models are linear, exponential and power law regressions for two variable data sets, also linear and power law fits for three variable data sets, i.e., two independent variables.

DATA-FIT allows users to generate graphical representations of their data with just two keystrokes. The graphics options include X-Y plots (scatter plots) of the experimental data, the best curves fitted to the experimental data and the confidence limits to which the regressions hold. The scales of the graphs automatically adjust to the data ranges. Users also have access to the LOTUS 1-2-3 graphics menu which allow them to tailor the graphs to their particular specifications, titles and data labels.

DATA-FIT contains file handling features which allows users to create data files, save their DATA-FIT results and even import data files from their LOTUS 1-2-3 spreadsheets.

DATA-FIT is menu driven so that the user does not have to memorize complicated commands to accomplish tasks. All the time-consuming keystrokes have been condensed into LOTUS 1-2-3 macros which are executed by just depressing two keys. The instructions that follow are based on an IBM or IBM-compatible machine. If you are using a WANG computer, the steps are the same, but the keystroke [ALT] is replaced by the [GL] key.

GETTING STARTED

In order to access the DATA-FIT program:

1 Boot up the LOTUS 1-2-3 spreadsheet from drive A;
2 Insert the DATA-FIT diskette into Drive B;
3 Retrieve the DATA-FIT menu file (FR "menu"). Refer to Figure A-1.

```
                        DATA-FIT MENU
DATA FILES          -----------------------------)[ALT][F]

TWO VARIABLES:
LINEAR REGRESSION   ---------------------)[ALT][L]
POWER LAW           ---------------------------)[ALT][P]
EXPONENTIAL         ---------------------------)[ALT][E]

THREE VARIABLES:
LINEAR REGRESSION   ---------------------)[ALT][V]
POWER LAW           ---------------------------)[ALT][Z]

QUIT  -----------------------------------------)[ALT][Q]
```

FIGURE A-1. Menu screen.

DATA FILES

The first step to initiate the DATA-FIT operation is to create a temporary data file from which your data will be read and processed by the respective models. You have two options. You can either enter the data manually, or, if you have LOTUS 1-2-3 data files on your personal data disk, DATA-FIT will automatically load it on to the data file for you.

To retrieve the DATA-FIT data file while in the main menu depress **[ALT] [F]** simultaneously (refer to Figure A-1).

Manual Input

When working with small data sets the quickest way to obtain your results is to enter the data manually on to the DATA-FIT file (called DATA1 File). Be sure to enter the data under the column headings to which they correspond. The X_1 and X_2 columns are for the independent variables and the Y column is for the dependent variable. If you choose a two variable regression, DATA-FIT will ignore the X_2 data column. After the data have been entered in the proper columns hit **[ALT] [M]** to choose a model from the main menu.

Importing a LOTUS 1-2-3 Data File

In order for DATA-FIT to incorporate the data set on to its data file, the worksheet range (wherein the data to be imported lies) must have been given a name. If the range was not previously named, remove the DATA-FIT diskette; insert your personal data disk; call up your LOTUS 1-2-3 worksheet; name the range by typing **/RNC**"**range name**" (enter) "**range**" (enter). After naming the data range be sure to re-save the LOTUS 1-2-3 worksheet file. Replace your data diskette with the DATA-FIT diskette and get back into the data file. Once in the data file hit **[ALT] [I]** to import your named data range. Position your data in the proper columns (see Figure A-2) and strike **[ALT] [D]**. This last instruction directs DATA-FIT to range/name/protect the data in the

```
                        X1      Y      X2
           ENTER       -----  -----  -----
           DATA-->     0.92   1.97    40
                        0.9   2.37    45
                       0.72   2.06    46
                       0.35   2.12    46
                        2.6    2.1    47
                        0.9    2.4    48
                       0.92   2.18    49
                        2.6   2.11    49
                        2.6   2.15    49
                       0.81   2.27   49.5
                       0.39   2.21    50
                       1.11   2.42    53
```

FIGURE A-2. DATA-FIT's data file (called DATA1).

X_1, X_2 and Y data columns. You may now save the data on a separate diskette (follow the DATA-FIT instructions on the program itself) or begin the regression. To begin the regression, hit **[ALT]** **[M]** to choose the models.

Caution: DO NOT import formula ranges. If the ranges which these formulas refer to are not also imported, DATA-FIT will not know what to do.

PERFORMING THE REGRESSION

As an example, we select the linear regression option for two variables. This regression gives the best straight line $Y = b_0 + b_1 X_1$. Depressing the **[ALT]**

```
     BEFORE YOU BEGIN TYPE IN THE Student t STATISTIC =    2.78
                                  Degrees of Freedom =       4
                                  Confidence          =   0.975
              move cursor down to see all results
 ** ******************RESULTS**************************
 **                  Y = (5190.239 )*X +-786.306    **
 **        6     = n                                **
 ** 0.065921   = ST. DEV. FOR X                     **
 **    5.733   = SUM OF X                           **
 **   25037.8  = SUM OF Y                           **
 ** 5.503955  = SUM OF X^2       ------------------  **
 **    1.1E+08 = SUM OF Y^2      HIT [ALT][O] FOR    **
 ** 24058.94   = SUM OF X*Y      OTHER OPTIONS       **
 **    0.9555  = X MEAN          ------------------  **
 ** 4172.966  = Y MEAN                              **
 ** 0.026073  = SUM X^2 - ((SUM OF X)^2)/N          **
 ** 992220.2  = SUM Y^2 - ((SUM OF Y)^2)/N          **
 ** 135.3277  = SUM X*Y - (SUM OF X)*(SUM OF Y)/N   **
 ** 5190.239  = SLOPE b0                            **
 ** -786.306  = INTERCEPT b1                        **
 ** 0.841362  = r* (COEFFICIENT OF CORRELATION)     **
 ** 269.1826  =ST.DEV. OF POINTS ABOUT REGRESSION   **
 ** Y RANGE:    3554.2     TO      4911.9           **
 ** X RANGE:    0.892      TO       1.081           **
 ** ************************************************ **
```

FIGURE A-3. Example results for two-variable linear regression.

[L] keys while in the main menu calls the two-variable linear regression subroutine. After a few seconds of processing the results screen appears.

The results of the two-variable linear regression include the slope of the best line b_1, the intercept b_0 and the coefficient of correlation r^*. See Figure A-3.

Table A-1. Percentile Values (t_p) for Student's t Distribution with ν Degrees of Freedom (Shaded Area = p).

ν	$t_{.995}$	$t_{.99}$	$t_{.975}$	$t_{.95}$	$t_{.90}$	$t_{.80}$	$t_{.75}$	$t_{.70}$	$t_{.60}$	$t_{.55}$
1	63.66	31.82	12.71	6.31	3.08	1.376	1.000	.727	.325	.158
2	9.92	6.96	4.30	2.92	1.89	1.061	.816	.617	.289	.142
3	5.84	4.54	3.18	2.35	1.64	.978	.765	.584	.277	.137
4	4.60	3.75	2.78	2.13	1.53	.941	.741	.569	.271	.134
5	4.03	3.36	2.57	2.02	1.48	.920	.727	.559	.267	.132
6	3.71	3.14	2.45	1.94	1.44	.906	.718	.553	.265	.131
7	3.50	3.00	2.36	1.90	1.42	.896	.711	.549	.263	.130
8	3.36	2.90	2.31	1.86	1.40	.889	.706	.546	.262	.130
9	3.25	2.82	2.26	1.83	1.38	.883	.703	.543	.261	.129
10	3.17	2.76	2.23	1.81	1.37	.879	.700	.542	.260	.129
11	3.11	2.72	2.20	1.80	1.36	.876	.697	.540	.260	.129
12	3.06	2.68	2.18	1.78	1.36	.873	.695	.539	.259	.128
13	3.01	2.65	2.16	1.77	1.35	.870	.694	.538	.259	.128
14	2.98	2.62	2.14	1.76	1.34	.868	.692	.537	.258	.128
15	2.95	2.60	2.13	1.75	1.34	.866	.691	.536	.258	.128
16	2.92	2.58	2.12	1.75	1.34	.865	.690	.535	.258	.128
17	2.90	2.57	2.11	1.74	1.33	.863	.689	.534	.257	.128
18	2.88	2.55	2.10	1.73	1.33	.862	.688	.534	.257	.127
19	2.86	2.54	2.09	1.73	1.33	.861	.688	.533	.257	.127
20	2.84	2.53	2.09	1.72	1.32	.860	.687	.533	.257	.127
21	2.83	2.52	2.08	1.72	1.32	.859	.686	.532	.257	.127
22	2.82	2.51	2.07	1.72	1.32	.858	.686	.532	.256	.127
23	2.81	2.50	2.07	1.71	1.32	.858	.685	.532	.256	.127
24	2.80	2.49	2.06	1.71	1.32	.857	.685	.531	.256	.127
25	2.79	2.48	2.06	1.71	1.32	.856	.684	.531	.256	.127
26	2.78	2.48	2.06	1.71	1.32	.856	.684	.531	.256	.127
27	2.77	2.47	2.05	1.70	1.31	.855	.684	.531	.256	.127
28	2.76	2.47	2.05	1.70	1.31	.855	.683	.530	.256	.127
29	2.76	2.46	2.04	1.70	1.31	.854	.683	.530	.256	.127
30	2.75	2.46	2.04	1.70	1.31	.854	.683	.530	.256	.127
40	2.70	2.42	2.02	1.68	1.30	.851	.681	.529	.255	.126
60	2.66	2.39	2.00	1.67	1.30	.848	.679	.527	.254	.126
120	2.62	2.36	1.98	1.66	1.29	.845	.677	.526	.254	.126
∞	2.58	2.33	1.96	1.645	1.28	.842	.674	.524	.253	.126

Source: R. A. Fisher and F. Yates, *Statistical Tables for Biological, Agricultural and Medical Research* (5th edition), Table III, Oliver and Boyd Ltd., Edinburgh.

```
        ********* OPTIONS *********
GRAPHICS
---------
        EXPERIMENTAL DATA POINTS ONLY     -------->[ALT][P]
        BEST LINE THROUGH DATA POINTS     -------->[ALT][L]
        BEST LINE PLUS CONFIDENCE LIMITS ------->[ALT][C]
        LOTUS GRAPHICS MENU -------------------->[ALT][G]
    DATA
    ---------
        FOR COLUMNS OF ALL
                CALCULATED DATA     -------->[ALT][D]
SAVE FILE
---------
        [ALT][S] "FILE NAME" [RETURN]

RESULTS----->[ALT][R]

MENU----->[ALT][M]            QUIT---->[ALT][Q]
```
FIGURE A-4. Options menu.

Before viewing all the results you have the option of quantifying the confidence of the regression through the use of the confidence limits. On the top of the Results screen, DATA-FIT asks for the Student t statistic for the probability of confidence based on the degrees of freedom for your data set. The degrees of freedom are automatically computed by DATA-FIT and are shown on the top of the screen. Use Table A-1 to obtain t for the degree of confidence and upon typing the values in, DATA-FIT will automatically compute the confidence intervals about the regression before you've finished viewing the Results box. For other options in the two variable linear regression, depress the [ALT][O] keys. This brings us to the screen OPTIONS menu, shown in Figure A-4.

OPTIONS

Main Features

The OPTIONS menu (see Figure A-4) includes access to columns of calculated data. It allows for various types of graphical representations and can save the DATA-FIT worksheet on the user's data diskette. In order to save the graphical representations for later printing, the LOTUS 1-2-3 graphics menu must be used.

The GRAPHICS OPTIONS will provide you with a scatter plot of the raw data (command [ALT] [P]), a plot of the best curve fit through the data points (command [ALT] [L]) and the best curve fit plus the confidence limits (command [ALT] [C]). See Figure A-5. Note that after a particular plot is generated on your screen, you are in the LOTUS 1-2-3 graphics menu. To view the next plot or to move to a new OPTION, hit Q to quit the graphics mode.

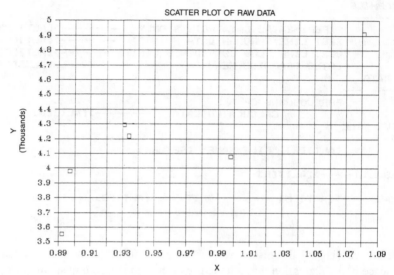

FIGURE A-5a. Plots provided by DATA-FIT—Scatter plot.

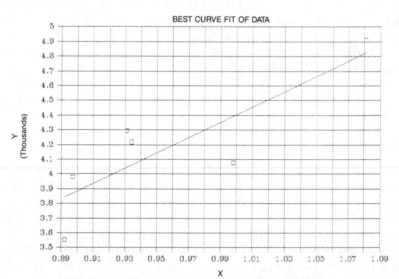

FIGURE A-5b. Plots provided by DATA-FIT—Best line fit.

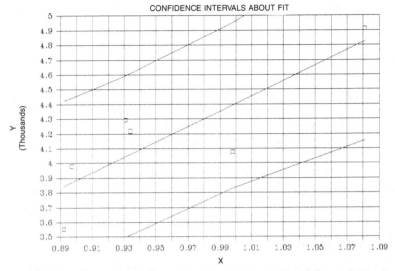

FIGURE A-5c. Plots provided by DATA-FIT—Best line fit with confidence limits.

The DATA OPTION (command **[ALT] [O]**) allows you to view all calculations including a point-by-point comparison of predicted and experimental values. Refer to Figure A-6. Command **[ALT] [R]** returns you to the RESULTS box, **[ALT] [M]** returns you to the main menu to choose a different regression model and **[ALT] [Q]** allows you to exit the program.

Other Features

Save file: To save your DATA-FIT regression results hit **[ALT] [S]** while in the options menu. DATA-FIT will prompt you to replace the program diskette with your personal data diskette. Then follow the LOTUS 1-2-3 menu and give your results file a name. Then DATA-FIT will tell you to re-insert the program diskette.

```
************CALCULATED DATA************

                       !-CONFIDENCE
                        LIMITS---!
    X        Y      pred Y    0.025    0.975    X^2        Y^2        X*Y
--------  --------  --------  --------  --------  --------  --------  --------
  0.892   3554.2   3843.386  2789.871  4896.901  0.795664  12632337  3170.346
  0.934   4217.7   4061.376  3063.938  5058.814  0.872356  17788993  3939.331
  0.931   4295.8   4045.805  3046.141  5045.469  0.866761  18453897  3999.389
  0.897   3979.9   3869.337  2825.190  4913.484  0.804609  15839604  3569.970
  0.998   4078.3   4393.551  3374.640  5412.463  0.996004  16632530  4070.143
  1.081   4911.9   4824.341  3604.749  6043.933  1.168561  24126761  5309.763
                0         0         0         0         0         0
                0         0         0         0         0         0
                0         0         0         0         0         0
                0         0         0         0         0         0
                0         0         0         0         0         0
```

FIGURE A-6. Calculated data screen for linear regression example.

Results: To return to the results screen hit **[ALT] [R]** while in the OPTIONS menu.

MENU: To choose a different model return to the main menu by hitting **[ALT] [M]**.

EXPONENTIAL AND POWER LAW REGRESSIONS

The exponential regression solves for the constants a and b in the equation $Y = ae^{bx}$. The power law does the same for data that can be modeled by the relation $Y = aX^b$.

The results and options features for these regressions are similar to those of the linear regression. There may be minor differences in the keystrokes. Nevertheless, the menus in DATA-FIT explain the differences.

THREE-VARIABLE REGRESSION MODELS

To begin the three-variable regression follow the instructions outlined in the DATA FILES section of this appendix. The only difference will be that now, in addition to the X_1 and Y columns you will also be filling in the X_2 column. Once you choose from the main menu either **[ALT] [V]** for a linear regression or **[ALT] [Z]** for a power law fit, DATA-FIT will automatically know that it has to include the X_2 data set in its calculations.

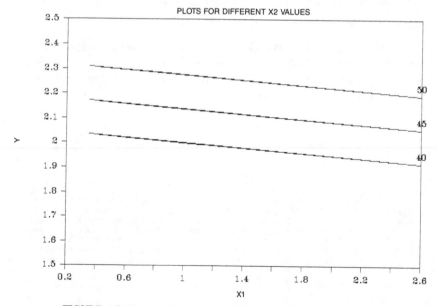

FIGURE A-7. Contour plot provided in three-variable regression routines.

POWER LAW—Choosing this model from the menu fits the three-variable set to the equation $Y = b_0 X_1^{b_1} X_2^{b_2}$.

LINEAR—The results of the linear regression with two independent variables yield the model $Y = b_0 + b_1 X_1 + b_2 X_2$.

RESULTS—The results of these two models differ from those of the two variable models in that the three-variable results give a correlation coefficient for both X_1 and X_2. Also included in the results are the constants b_0, b_1 and b_2.

OPTIONS—In the options menu of the three-variable regressions the only distinct feature is the graphics. Since the LOTUS 1-2-3 release 1A does not allow three-dimensional plots, the DATA-FIT graphics feature for the regression on three-variable data sets yields three contour plots of Y vs X_1 at three values of X_2 which you specify. See Figure A-7.

You also have access to the LOTUS graphics menu to incorporate other options to the graph.

To save your DATA-FIT three-variable regression results hit [ALT] [S] while in the options menu. DATA-FIT will tell you to replace the program diskette with your personal data diskette. Then follow the LOTUS 1-2-3 menu and give your results file a name. Then DATA-FIT will tell you to re-insert the program diskette.

To choose a different model return to the main menu by hitting [ALT] [M].

IMPORTANT: If you ever find yourself lost in a strange part of the LOTUS worksheet while using one of the DATA-FIT models, hit [ALT] [O]. This will return you to the options menu.

QUITTING: When you are done fitting your data to the desired models and have saved all your work, return to the main menu and hit [ALT] [Q] to quit.

General Definitions and Miscellaneous Statistical Methods

NESTED DESIGNS

Nested or hierarchal designs are based on a definite relationship between two factors. In this relationship, every level of Factor B appears with only one level of Factor A. This situation is characterized as Factor B being nested in Factor A. As an example, consider the following response and factors:

RESPONSE — Green Strength of Compound Polymer
FACTOR A — Carbon Black (e.g., several supplies used)
FACTOR B — Oil (several suppliers used)

The layout of the experimental design might take the form:

	Carbon Black		
	I	II	III
Oils	1	4	7
	2	5	8
	3	6	9

The factors can be described as being *crossed*, since all levels of Factor B appear with Factor A in a full factorial design. We can estimate the components of variance, where the overall relationship is:

$$\sigma^2 = \sigma_A^2 + \sigma_B^2 + \sigma_C^2 + \ldots$$

In the above illustration, we are interested in understanding why there is a variation in the compound green strength. The components of variance would therefore be assigned such that σ_A^2 is the carbon black variability, σ_B^2 is the oil variability (e.g., oil viscosity or purity level), and σ_C^2 is the test variability.

This type of analysis is sometimes referred to as a hierarchal design. The reason for this terminology is that Factor B is nested in Factor A, and in turn Factor C is in B, and so forth. The design establishes the number of nested levels of each factor n_A, n_B, n_C . . . Figure B-1 shows the scheme for a three-factor nested design for illustration. As a general rule, to normalize the

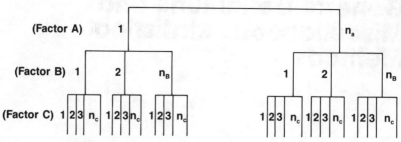

FIGURE B-1. Illustrates a three-factor nested design scheme.

degrees of freedom for each factor, the number of levels for all but Factor A should be kept at two.

For a two-factor nested design (i.e., two components of variance), Factor B levels are essentially replicates of Factor A. Another way of viewing this is that Factor B is a random variable. The response model is:

$$Y_{ij} = \mu + A_i + B_{j(i)}$$

where Y_{ij} = response at the ith level of Factor A and the jth nested condition of Factor B.

μ = overall response mean.

A_i = random variable with mean zero and variance σ_A^2 (where $i = 1, \ldots n_A$).

$B_{u(i)}$ = random variable with mean zero and variance of σ_B^2 (where $j = 1, \ldots n_B$).

The response variation is thus:

$$\sigma_y^2 = \sigma_A^2 + \sigma_B^2$$

The standard deviations squared S_A^2 and S_B^2 must be estimated for σ_A^2 and σ_B^2, respectively.

The average response $\bar{Y}$ can be estimated from n_A levels of Factor A with n_B levels of Factor B nested in each level of Factor A. The variance of $\bar{Y}$ can be computed from:

$$S_{\bar{y}}^2 = \frac{S_A^2}{n_A} + \frac{S_B^2}{n_A n_B}$$

The ANOVA scheme described in the text can be used to evaluate this type of problem.

For three components of variance (i.e., analysis of three-factor nested design), Factor C levels are nested in levels of Factor B, and Factor B levels are nested in levels of Factor A. That is, all three factors are essentially random variables in the components of the variance model. The response model is:

$$Y_{ijk} = \mu + A_i + B_{j(i)} + C_{k(ij)}$$

where Y_{ijk} = response at the ith level of A, j^{th} level of B and at the kth level of C.

μ = overall response mean.

A_i, $B_{j(i)}$, $C_{k(ij)}$ = random variables with mean zero and variances σ_A^2, σ_B^2, σ_C^2, respectively.

The subscripts i, j and k cover the run ranges $1 \ldots n_A$, $1 \ldots n_B$, and $1 \ldots n_C$, respectively.

For average response $\bar{Y}$ estimated from n_A levels of A, n_B levels of B, and n_C levels of C, the variance is:

$$S_{\bar{y}}^2 = \frac{S_A^2}{n_A} \frac{S_B^2}{n_A n_B} + \frac{S_C^2}{n_A n_B n_C}$$

PROBIT ANALYSIS

This is a technique for assessing dose-response relationships. The response in this problem class is quantal (i.e., all-or-nothing). The probit model is based on the assumption that the tolerance distribution is Gaussian in either dose or log (dose), see Figure B-2.

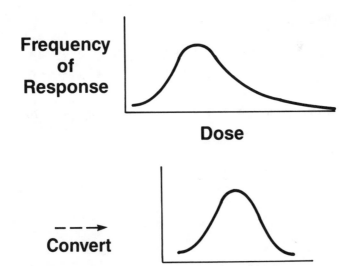

FIGURE B-2. Response must be normal to either dose or log (dose).

The design is formulated such that several doses are chosen with the 50% response dose bracketed. If the 50% dose is not known, divide the N subjects equally among the selected doses. For large N, it's possible to run a pilot experiment with $N/10$ subjects to obtain an approximation of the 50% dosage. The interval of doses should be wide enough such that the extremes include the 5–15% and the 85–95% response ranges.

The analysis scheme is a weighted regression, where the highest weighting occurs for responses close to the 50% response and tails off in either direction. The estimation procedure is iterative (described in standard statistics textbooks), where doses giving 0 or 100% response are ignored in the scheme.

TRIMMING

This is a technique aimed at eliminating the extreme values at each end of a sample. This method is applied when individuals in the sample are suspected of having an origin from a different population. In this case, we may want to estimate the mean in such a manner so as to minimize the effects of these extraneous data. Another similar method is *Wintorization*. In this procedure, the smallest and largest observations are assigned the value of their closest neighbors. Both are illustrated by the sample data set:

$$X_{(1)} = 67.9 \qquad X_{(3)} = 73.7 \qquad X_{(5)} = 85.5$$

$$X_{(2)} = 68.2 \qquad X_{(4)} = 78.6 \qquad X_{(6)} = 85.9$$

$$\bar{X} = 459.8/6 = 76.63$$

For trimming:

$$\bar{X} = (68.2 + 73.7 + 78.6 + 85.5)/4 = 76.50$$

For Wintorization:

$$\bar{X} = (68.2 + 68.2 + 73.7 + 78.6 + 85.5 + 85.5)/6 = 76.62$$

BIBLIOGRAPHY

ATCHLEY, W. R. and E. H. BRYANT, eds. *Multivariate Statistical Methods, Volume 1—Among-Groups Covariation.* Dowden, Hutchinson & Ross Pub., Inc., Stroudsburg, PA (1975).

AZBEL, D. S. and N. P. CHEREMISINOFF. *Fluid Mechanics and Unit Operations.* Ann Arbor Science Pub., Ann Arbor, MI (1983).

BENDAT, J. S. *Principles and Applications of Random Noise Theory.* John Wiley & Sons Pub., New York (1977).

BENDAT, J. S. and A. G. PIERSOL. *Engineering Applications of Correlation and Spectral Analysis.* John Wiley & Sons Pub., New York (1980).

BENDAT, J. S. and A. G. PIERSOL. *Random Data Analysis and Measurement Procedures.* Wiley-Interscience, New York (1971).

BRYANT, E. H. and W. R. ATCHLEY, eds. *Multivariate Statistical Methods: Volume 2—With-Groups Covariation.* Dowden, Hutchinson & Ross Pub., Inc., Stroudsburg, PA (1975).

CHEREMISINOFF, N. P. *Instrumentation for Complex Fluid Flows.* Technomic Publishing Co., Lancaster, PA (1986).

CHEREMISINOFF, N. P. and P. N. CHEREMISINOFF. *Hydrodynamics of Gas-Solids Fluidization.* Gulf Publishing Co., Houston, TX (1984).

FISCHER, R. A. *Statistical Methods for Research Workers.* Oliver and Boyd Pub., Edinburgh, England (1950).

FISCHER, R. A. and F. YATES. *Statistical Tables.* 6th ed., Oliver and Boyd (1967).

GORMAN, J. W. and J. E. HINMAN, "Simplex Lattice Designs for Multicomponent Systems," *Technometrics, 4*(4), 463–487 (Nov. 1962).

JOHNSON, N. L. and F. C. LEONE, *Industrial Quality Control,* pp. 15–21 (June 1962).

JOHNSON, N. L. and F. C. LEONE, *Industrial Quality Control,* pp. 29–36 (July 1962).

JOHNSON, N. L. and F. C. LEONE, *Industrial Quality Control,* pp. 22–28 (August 1962).

LUCAS, J., *Journal of Quality Technology,* pp. 1–12 (Jan. 1976).

MIDDLEBROOKS, E. J., *Statistical Computations: How to Solve Statistical Problems.* Ann Arbor Science Pub. Inc., Ann Arbor, MI (1976).

MILLER, I. and J. E. FREUND. *Probability and Statistics for Engineers.* Prentice-Hall, Inc., Englewood Cliffs, NJ (1965).

MILLER, R. E., "Analysis of Variance," *Chemical Engineering Magazine, 92*(6), 173–178 (March 18, 1985).

NARCY, J. P. and J. RENAUD, *Journal of the American Oil Chem. Soc., 49*(10), 598–608 (1972).

PLACKETT, R. L. and J. P. BURMAN, *Biometrika, 33,* 305–325 (1946).

SCHEFFE, H., *Statist. Soc. B.*, *20*, 344 (1958).

SCHEFFE, H., *Ibid*. *25*, 235 (1963).

SNEDECOR, G. W. and W. G. COCHRAN, *Statistical Methods*. Seventh edition, Iowa State University Press, Ames, IA (1980).

WALPOLE, R. E. and R. H. MYERS. *Probability and Statistics for Engineers and Scientists*. MacMillan Pub. Co., New York (1972).

SUGGESTED READINGS

COCHRAN, W. G. and G. M. COX. *Experimental Designs.* John Wiley & Sons, New York (1957).

DAVIES, O. L. *The Design and Analysis of Industrial Experiments.* Hafner Publishing Co., New York (1967).

DIXON, W. J. and F. J. MASSEY. *Introduction to Statistical Analysis.* McGraw-Hill Book Co., Inc., New York (1969).

DRAPER, N. R. and H. SMITH. *Applied Regression Analysis.* John Wiley & Sons, New York (1966).

DUNCAN, A. J. *Quality Control and Industrial Statistics.* Richard D. Irwin, Inc., Homewood, IL (1965).

MANDEL, J., *The Statistical Analysis of Experimental Data.* Interscience Publishers, New York (1964).

INDEX

205

☐ YES, send me the new software support package that will alleviate the difficult, time consuming calculations of both linear and non-linear regressions with up to three variable inputs. I understand that DATA·FIT is IBM and Wang machine compatible and requires approximately 300 K of memory. The program is designed as a subroutine using the popular Lotus 1-2-3 as a master program. In addition to calculating the best curve fit through a set of data points, DATA·FIT also has the ability to provide a variety of graphical representations for the purpose of data interpretation. ONLY **$45.**

ORDER FORM: **DATA·FIT** ISBN: 87762-506-9

$45.00

ORGANIZATION _____

NAME_____

ADDRESS _____

CITY_____STATE_____ZIP _____

CREDIT CARD ORDERS: ☐ VISA   ☐ Mastercard

CARD NO._____EXP. DATE _____

SIGNATURE _____

Toll-Free 800 Number For Fast Delivery on Credit Card Orders
To order with Visa or Mastercard, phone
Outside PA: **1-800-233-9936**   In PA: **(717) 291-5609**
This is a direct line to our order desk for all credit card orders.
From 8:30 A.M. to 5:00 P.M. Eastern Time

Please detach and return to:
TECHNOMIC Publishing Company, Inc.
851 New Holland Ave., Box 3535, Lancaster, PA 17604, U.S.A.